THE SCIENCE OF INTERSTELLAR

ALSO BY KIP THORNE

The Future of Spacetime (with Stephen W. Hawking, Igor Novikov, Timothy Ferris, Alan Lightman, and Richard H. Price)

Black Holes & Time Warps: Einstein's Outrageous Legacy

Black Holes: The Membrane Paradigm (with Richard H. Price and Douglas A. MacDonald)

Gravitation (with Charles W. Misner and John Archibald Wheeler)

Gravitation Theory and Gravitational Collapse (with B. Kent Harrison, Masami Wakano, and John Archibald Wheeler)

THE SCIENCE OF INTERSTELLAR

KIP THORNE

W. W. NORTON & COMPANY
New York • London

Printed in the United States of America
First Edition

For information about permission to reproduce selections from this book,
write to Permissions, W. W. Norton & Company, Inc.,
500 Fifth Avenue, New York, NY 10110

For information about special discounts for bulk purchases, please contact
W. W. Norton Special Sales at specialsales@wwnorton.com or 800-233-4830

Manufacturing by Courier Kendallville
Book design by Chris Welch
Production managers: Julia Druskin and Joe Lops

ISBN 978-0-393-35137-8

W. W. Norton & Company, Inc.
500 Fifth Avenue, New York, N.Y. 10110
www.wwnorton.com

W. W. Norton & Company Ltd.
Castle House, 75/76 Wells Street, London W1T 3QT

3 4 5 6 7 8 9 0

CONTENTS

I. FOUNDATIONS

II. GARGANTUA

III. DISASTER ON EARTH

IV. THE WORMHOLE

V. EXPLORING GARGANTUA'S ENVIRONS

VI. EXTREME PHYSICS

VII. CLIMAX

FOREWORD

One of the great pleasures of working on *Interstellar* has been getting to know Kip Thorne. His infectious enthusiasm for science was obvious from our first conversation, as was his reluctance to proffer half-formed opinions. His approach to all the narrative challenges that I threw him was always calm, measured and above all, *scientific.* In trying to keep me on the path of plausibility, he never showed impatience with my unwillingness to accept things on trust (although my two-week challenge to his faster-than-light prohibition might have elicited a gentle sigh).

He saw his role not as science police, but as narrative collaborator—scouring scientific journals and academic papers for solutions to corners I'd written myself into. Kip has taught me the defining characteristic of science—its humility in the face of nature's surprises. This attitude allowed him to enjoy the possibilities that speculative fiction presented for attacking paradox and unknowability from a different angle—storytelling. This book is ample demonstration of Kip's lively imagination and his relentless drive to make science accessible to those of us not possessed of his massive intellect or his immense body of knowledge. He wants people to understand and get excited about the crazy truths of our universe. This book is structured to let the reader dip in to a topic as deeply as their affinity for science prompts them—no one is left behind, and everyone gets to experience some of the fun I had trying to keep up with Kip's agile mind.

Christopher Nolan
Los Angeles, California
July 29, 2014

PREFACE

I've had a half-century-long career as a scientist. It's been joyously fun (most of the time), and has given me a powerful perspective on our world and the universe.

As a child and later as a teenager, I was motivated to become a scientist by reading science fiction by Isaac Asimov, Robert Heinlein, and others, and popular science books by Asimov and the physicist George Gamow. To them I owe so much. I've long wanted to repay that debt by passing their message on to the next generation; by enticing youths and adults alike into the world of science, real science; by explaining to non-scientists how science works, and what great power it brings to us as individuals, to our civilization, and to the human race.

Christopher Nolan's film *Interstellar* is an ideal messenger for that. I had the great luck (and it *was* luck) to be involved with *Interstellar* from its inception. I helped Nolan and others weave real science into the film's fabric.

Much of *Interstellar's* science is at or just beyond today's frontiers of human understanding. This adds to the film's mystique, and it gives me an opportunity to explain the differences between firm science, educated guesses, and speculation. It lets me describe how scientists take ideas that begin as speculation, and prove them wrong or transform them into educated guesses or firm science.

I do this in two ways: First, I *explain* what is known today about phenomena seen in the movie (black holes, wormholes, singularities, the fifth dimension, and the like), and I explain how we learned what we know, and how we hope to master the unknown. Second, I *interpret*, from a scientist's viewpoint, what we see in *Interstellar*, much as an art critic or ordinary viewer interprets a Picasso painting.

My interpretation is often a description of what I imagine might be going on behind the scenes: the physics of the black hole Gargantua, its singularities, horizon, and visual appearance; how Gargantua's tidal gravity could generate 4000-foot water waves on Miller's planet; how the tesseract, an object with four space dimensions, could transport three-dimensional Cooper through the five-dimensional bulk; . . .

Sometimes my interpretation is an *extrapolation* of *Interstellar*'s story beyond what we see in the movie; for example, how Professor Brand, long before the movie began, might have discovered the wormhole, via gravitational waves that traveled from a neutron star near Gargantua through the wormhole to Earth.

These interpretations, of course, are my own. They are not endorsed by Christopher Nolan any more than an art critic's interpretations were endorsed by Pablo Picasso. They are my vehicle for describing some wonderful science.

Some segments of this book may be rough going. That's the nature of real science. It requires thought. Sometimes deep thought. But thinking can be rewarding. You can just skip the rough parts, or you can struggle to understand. If your struggle is fruitless, then that's my fault, not yours, and I apologize.

I hope that at least once you find yourself, in the dead of night, half asleep, puzzling over something I have written, as I puzzled at night over questions that Christopher Nolan asked me when he was perfecting his screenplay. And I especially hope that, at least once in the dead of night, as you puzzle, you experience a Eureka moment, as I often did with Nolan's questions.

I'm grateful to Christopher Nolan, Jonathan Nolan, Emma Thomas, Lynda Obst, and Steven Spielberg for welcoming me into Hollywood, and giving me this wonderful opportunity to fulfill my dream, to pass on to the next generation my message of the beauty, the fascination, and the power of science.

Kip Thorne
Pasadena, California
May 15, 2014

THE SCIENCE OF
INTERSTELLAR

1

A Scientist in Hollywood:

THE GENESIS OF *INTERSTELLAR*

Lynda Obst, My Hollywood Partner

The seed for *Interstellar* was a failed romance that warped into a creative friendship and partnership.

In September 1980, my friend Carl Sagan phoned me. He knew I was a single father, raising a teenaged daughter (or trying to do so; I wasn't very good at it), and living a Southern California single's life (I was only a bit better at that), while pursuing a theoretical physics career (at *that* I was a lot better).

Carl called to propose a blind date. A date with Lynda Obst to attend the world premier of Carl's forthcoming television series, *Cosmos*.

Lynda, a brilliant and beautiful counterculture-and-science editor for the *New York Times Magazine*, was recently transplanted to Los Angeles. She had been dragged there kicking and screaming by her husband, which contributed to their separation. Making the best of a seemingly bad situation, Lynda was trying to break into the movie business by formulating the concepts for a movie called *Flashdance*.

The *Cosmos* premier was a black-tie event at the Griffith Observatory. Klutz that I was, I wore a baby-blue tuxedo. Everybody who was anybody in Los Angeles was there. I was completely out of my element, and had a glorious time.

For the next two years, Lynda and I dated on and off. But the chemistry just wasn't right. Her intensity enthralled and exhausted me. I debated whether the exhaustion was worth the highs, but the choice wasn't mine. Perhaps it was my velour shirts and double-knit pants; I don't know. Lynda soon lost romantic interest in me, but something better was growing: a lasting and creative friendship and partnership between two very different people, from very different worlds.

Fast-forward to October 2005, another of our occasional one-on-one dinners, where conversation would range from recent cosmological discoveries, to left-wing politics, to great food, to the shifting sands of moviemaking. Lynda by now was among Hollywood's most accomplished and versatile producers (*Flashdance, The Fisher King, Contact, How to Lose a Guy in Ten Days*). I had married. My wife, Carolee Winstein, had become best friends with Lynda. And I'd not done badly in the world of physics.

Over dinner, Lynda described an idea she had conceived for a science-fiction movie and asked me to help her flesh it out. This would be her second venture into science fiction: a collaboration with me, modeled on her previous collaboration with Carl Sagan on the movie *Contact*.

I never imagined myself helping create a movie. I never coveted a presence in Hollywood, beyond a vicarious one, through Lynda's adventures. But working with Lynda appealed to me, and her ideas involved wormholes, an astrophysics concept I had pioneered. So she easily lured me into brainstorming with her.

During the next four months, over a few dinners and e-mails and phone calls, we formulated a rough vision for the film. It included wormholes, black holes, and gravitational waves, a universe with five dimensions, and human encounters with higher-dimensional creatures.

But most important to me was our vision for a blockbuster movie *grounded from the outset in real science*. Science at and just beyond the frontiers of human knowledge. A film in which the director, screenwriters, and producers respect the science, take inspiration from it, and weave it into the movie's fabric, thoroughly and compellingly. A film that gives the audience a taste of the wondrous things that the laws of physics can and might create in our universe, and the great things humans can achieve by mastering the physical laws. A film that inspires many in the audience to go learn about the science, and perhaps even pursue careers in science.

Nine years later, *Interstellar* is achieving all we envisioned. But the path from there to here has been a bit like the "Perils of Pauline," with many a spot where our dream could have collapsed. We acquired and then lost the legendary director Steven Spielberg. We acquired a superb young screenwriter, Jonathan Nolan, and then lost him twice, at crucial stages, for many months each. The movie sat in limbo, directorless, for two and a half years. Then, wondrously, it was resurrected and transformed in the hands of Jonathan's brother, Christopher Nolan, the greatest director of his young generation.

Steven Spielberg, the Initial Director

In February 2006, four months after we began brainstorming, Lynda had lunch with Todd Feldman, Spielberg's agent at the Creative Artists Agency, CAA. When Feldman asked what movies she was working on, she described her collaboration with me, and our vision for a sci-fi movie with real science woven in from the outset—our dream for *Interstellar.* Feldman got excited. He thought Spielberg might be interested and urged Lynda to send him a treatment *that very day!* (A "treatment" is a description of the story and characters, usually twenty pages or longer.)

All we had in writing were a few e-mail exchanges and notes from a few dinner conversations. So we worked at whirlwind speed for a couple of days to craft an eight-paged treatment we were proud of, and sent it off. A few days later Lynda e-mailed me: "Spielberg has read it and is very interested. We may need to have a little meeting with him. Game? XX Lynda."

Of course I was game! But a week later, before any meeting could be arranged, Lynda phoned: "Spielberg is signing on to direct our *Interstellar!*" Lynda was ecstatic. I was ecstatic. "This kind of thing never happens in Hollywood," she told me. "Never." But it did.

I then confessed to Lynda that I had seen only one Spielberg movie in my life—*ET,* of course. (As an adult, I had never been all that interested in movies.) So she gave me a homework assignment: Spielberg Movies Kip Must Watch.

A month later, on March 27, 2006, we had our first meeting with Spielberg—or Steven, as I began to call him. We met in a homey conference room in the heart of his movie production company Amblin, in Burbank.

At our meeting, I suggested to Steven and Lynda two guidelines for the science of *Interstellar*:

1. Nothing in the film will violate firmly established laws of physics, or our firmly established knowledge of the universe.
2. Speculations (often wild) about ill-understood physical laws and the universe will spring from real science, from ideas that at least some "respectable" scientists regard as possible.

Steven seemed to buy in, and then accepted Lynda's proposal to convene a group of scientists to brainstorm with us, an *Interstellar* Science Workshop.

The workshop was on June 2 at the California Institute of Technology (Caltech), in a conference room down the hall from my office.

It was an eight-hour, free-wheeling, intoxicating discussion among fourteen scientists (astrobiologists, planetary scientists, theoretical physicists, cosmologists, psychologists, and a space-policy expert) plus Lynda, Steven, and Steven's father Arnold, and me. We emerged, exhausted but exhilarated with a plethora of new ideas and objections to our old ideas. Stimuli for Lynda and me, as we revised and expanded our treatment.

It took us six months due to our other commitments, but by January 2007 our treatment had grown to thirty-seven pages, plus sixteen pages about the science of *Interstellar*.

Jonathan Nolan, the Screenwriter

In parallel, Lynda and Steven were interviewing potential screenwriters. It was a long process that ultimately converged on Jonathan Nolan, a thirty-one-year-old who had coauthored (with his brother Christopher) just two screenplays, *The Prestige* and *The Dark Knight*, both big hits.

Jonathan, or Jonah as his friends call him, had little knowledge of science, but he was brilliant and curious and eager to learn. He spent many months devouring books about all the science relevant to *Interstellar* and asking probing questions. And he brought to our film big new ideas that Steven, Lynda, and I embraced.

Jonah was wonderful to work with. He and I brainstormed together many times about the science of *Interstellar,* usually over a two- or three-hour lunch at the Caltech faculty club, the Athenaeum. Jonah would come to lunch brimming with new ideas and questions. I would react on the spot: this is scientifically possible, that isn't, . . . My reactions were sometimes wrong. Jonah would press me: Why? What about . . . ? But I'm slow. I would go home and sleep on it. In the middle of the night, with my gut reactions suppressed, I would often find some way to make what he wanted to work, work. Or find an alternative that achieved the end he sought. I got good at creative thinking when half asleep.

The next morning, I would assemble the semicoherent notes I had written during the night, decipher them, and write Jonah an e-mail. He would respond by phone or e-mail or another lunch, and we would converge. In this way we came to gravitational anomalies, for example, and the challenge of harnessing them to lift humanity off Earth. And I discovered ways, just beyond the bounds of current knowledge, to make the anomalies scientifically possible.

At crucial times we brought Lynda into the mix. She was great at critiquing our ideas and would send us spinning off in new directions. In parallel with our brainstorming, she was working her magic to keep Paramount Pictures at bay so we could maintain our creative autonomy, and planning the next phases of turning *Interstellar* into a real movie.

By November 2007, Jonah, Lynda, Steven, and I had agreed on the structure for a radically revised story based on Lynda's and my original treatment, Jonah's big ideas, and the many other ideas that arose from our discussions—and Jonah was deep into writing. Then, on November 5, 2007, the Writers Guild of America called a strike. Jonah was forbidden to continue writing, and disappeared.

I panicked. Will all our hard work, all our dreams, be for naught? I asked Lynda. She counseled patience, but was clearly very upset. She vividly tells the story of the strike in scene 6 of her book *Sleepless in Hollywood*. The scene is titled "The Catastrophe."

The strike lasted three months. On February 12, when it ended, Jonah returned to writing and to intense discussions with Lynda and me. Over the next sixteen months, he produced a long, detailed outline for the screenplay, and then three successive drafts of the screenplay itself. When each was finished, we met with Steven to discuss it. Steven would ask probing questions for an hour or more before proffering suggestions, requests, or instructions for changes. He was not very hands-on, but he was thoughtful, incisive, creative—and sometimes firm.

In June 2009, Jonah gave Steven draft 3 of the screenplay, and disappeared from the scene. He had long ago committed to write *The Dark Knight Rises*, and had been delaying for month after month while working on *Interstellar*. He could delay no more, and we were without a screenwriter. On top of that, Jonah's father became gravely ill. Jonah spent many months in London by his father's side, until his father's death in December. Through this long hiatus, I feared that Steven would lose interest.

But Steven hung in there with us, awaiting Jonah's return. He and Lynda could have hired somebody else to complete the screenplay, but they so valued Jonah's talents that they waited.

Finally in February 2010 Jonah was back, and on March 3, Steven, Lynda, Jonah, and I had a very productive meeting to discuss Jonah's nine-month-old draft 3. I was feeling a bit giddy. At last we were back on track.

Fig. 1.1. Jonah Nolan, Kip, and Lynda Obst.

Then on June 9, with Jonah deep into draft 4, I got an e-mail from Lynda: "We have a Steven deal problem. I'm into it." But it was not soluble. Spielberg and Paramount could not reach an agreement for the next phase of *Interstellar,* and Lynda couldn't broker a solution. Suddenly we had no director.

Interstellar was going to be very expensive, Steven and Lynda had independently told me. There were very few directors with whom Paramount would entrust a movie of this magnitude. I envisioned *Interstellar* in limbo, dying a slow death. I was devastated. So was Lynda, at first. But she is a superb problem solver.

Christopher Nolan, the Director and Screenwriter

Only thirteen days after Lynda's we-have-a-Steven-deal-problem e-mail, I opened my e-mail queue to find a euphoric follow-on message: "Great talk with Emma Thomas . . ." Emma is Christopher Nolan's wife/producer and collaborator on all his movies. She and Christopher were interested. Lynda was tremulous with excitement. Jonah called and told her, "This is the best possible outcome." But the deal, for many reasons, would not be finalized for two and a half years, though we were fairly certain Christopher and Emma were committed.

So we sat. And waited. June 2010, through 2011, to September 2012. Throughout, I fretted. In front of me, Lynda projected an air of confidence. But she later confided having written these words to herself: "Tomorrow we could wake up and Chris Nolan could be gone, after two and a half years of waiting. He could come up with his own idea. Another producer could hand him a script he likes more. He could decide to take a break. Then I would have been wrong to have waited for him all this time. It happens. That is my life, the lives of creative producers. But he's the perfect director for us. So we wait."

At last negotiations began, far, far above my pay grade. Christopher Nolan would direct only if Paramount would share the movie with Warner Bros., the studio that had made his last few movies, so a deal—an extremely complex deal—had to be struck between the two studios, normally rivals.

Finally, on December 18, 2012, Lynda e-mailed: "par and warners agreed to terms. Well chop my liver! starting in spring!!!" And from then on, with *Interstellar* in Christopher Nolan's hands, so far as I could tell all was clear sailing. At last! Clear, fun, and invigorating.

Christopher knew Jonah's screenplay well. They are brothers, after all, and had talked as Jonah wrote. They have a phenomenally successful history of collaborating on screenplays: *The Prestige, The Dark Knight, The Dark Knight Rises*. Jonah writes the initial drafts, and then Christopher takes over and rewrites, thinking carefully about how he will film each scene as he crafts it on paper.

With *Interstellar* now fully in Christopher's own hands, he combined Jonah's script with the script from another project he'd been working on, and he injected a radically fresh perspective and a set of major new ideas—ideas that would take the movie in unexpected new directions.

In mid-January, Chris, as I soon came to call him, asked to meet me one-on-one in his office at Syncopy, his movie production company on the Warner Bros. lot.

As we talked, it became clear that Chris knew a remarkable amount of relevant science and had deep intuition about it. His intuition was occasionally off the mark, but usually right on. And he was tremendously curious. Our conversations often diverged from *Interstellar* to some irrelevant science issue that fascinated him.

In that first meeting, I laid on Chris my proposed science guidelines: Nothing will violate firmly established laws of physics; speculations will all spring from science. He seemed positively inclined, but told me that if I didn't like what he did with the science, I didn't have to defend him in public. That shook me up a bit. But with the movie now in postproduction, I'm impressed how well he followed those guidelines, while making sure they didn't get in the way of making a great movie.

Chris worked intensely from mid-January to early May rewriting Jonah's screenplay. From time to time he or his assistant, Andy Thompson, would phone me and ask that I come to his office or his home to talk about science issues, or come to read a new draft of his screenplay and then meet to discuss it. Our discussions were long, typically ninety minutes, sometimes followed by long phone calls a day or two later. He raised issues that made me think. As when working with Jonah, my best thinking was in the dead of night. The next morning I would write up my thoughts in a several-paged memo with diagrams and pictures, and hand carry them to Chris. (Chris worried about our ideas leaking out and spoiling his fans' mounting anticipation. He's one of the most secretive filmmakers in Hollywood.)

Chris's ideas occasionally seemed to violate my guidelines but, amazingly, I almost always found a way to make them work, scientifically. Only once did I fail miserably. In response, after several discussions over a two-week period, Chris backed off and took that bit of the film in another direction.

So in the end I have no qualms about defending what Chris did with the science. On the contrary, I'm enthusiastic! He turned into reality Lynda's and my dream of a blockbuster movie with foundations of real science, and with real science woven throughout its fabric.

In the hands of Jonah and Chris, *Interstellar*'s story changed enormously. It resembles Lynda's and my treatment only in broadest brushstrokes. It is so much better! And as for the science ideas: they are not all mine by any means. Chris brought remarkable science ideas of his own to the film, ideas that my physicist colleagues will assume were mine, ideas that I said to myself, when I saw them, Why didn't I think of that? And remarkable ideas arose from my discussions with Chris, with Jonah, and with Lynda.

Fig. 1.2. Kip and Christopher Nolan talking on set in the *Endurance*'s control module.

One April evening, Carolee and I threw a big party for Stephen Hawking at our home in Pasadena, with a diverse crowd of a hundred people: scientists, artists, writers, photographers, filmmakers, historians, schoolteachers, community organizers, labor organizers, business entrepreneurs, architects, and more. Chris and Emma came, as well as Jonah Nolan and his wife Lisa Joy, and of course Lynda. In the late evening, we stood together for a long time on a balcony, under the stars, far from the party noise, talking quietly—my first opportunity to get to know Chris as a man, rather than a filmmaker. It was so enjoyable!

Chris is down to earth, fascinating to talk with, and has a great sense of wry humor. He reminds me of another friend of mine, Gordon Moore, the founder of Intel: Both, at the pinnacle of their fields, completely unpretentious. Both driving old cars, preferring them to their other, more luxurious cars. Both making me feel comfortable and, introvert that I am, that's not easy.

Paul Franklin, Oliver James, Eugénie von Tunzelmann: The Visual-Effects Team

One day in mid-May 2013 Chris phoned me. He wanted to send a guy named Paul Franklin over to my home to discuss the computer graphics for *Interstellar.* Paul came the next day, and we spent a delightful two hours brainstorming in my home office. He was modest in demeanor, by contrast with Chris's forcefulness. He was brilliant. He showed a deep knowledge of the relevant science, despite having majored in the arts in college.

As Paul was leaving, I asked him which graphics company he was thinking of using for the visual effects. "Mine," he responded, mildly. "And what company is that?" I asked, naively. "Double Negative. We have 1000 employees in London and 200 in Singapore."

After Paul departed I Googled him and discovered that not only had he cofounded Double Negative, he had also won an Academy Award for the visual effects in Chris's movie *Inception.* "It's time I get educated about this movie business," I murmured to myself.

In a video conference a few weeks later, Paul introduced me to the London-based leaders of his *Interstellar* visual-effects team. Most rele-

Fig. 1.3. Paul Franklin and Kip.

vant to me were Oliver James, the chief scientist who would write computer code underlying the visual effects; and Eugénie von Tunzelmann, who led the artistic team that would take Oliver's computer code and add extensive artistic twists to produce compelling images for the movie.

Oliver and Eugénie were the first people with physics training that I had met on *Interstellar*. Oliver has a degree in optics and atomic physics, and knows the technical details of Einstein's special relativity. Eugénie is an engineer, trained at Oxford, with a focus on data engineering and computer science. They speak my language.

We quickly developed a great working relationship. For several months, I struggled near full time, formulating equations for images of the universe near black holes and wormholes (Chapters 8 and 15). I tested my equations using low-resolution, user-friendly computer soft-

Fig. 1.4. Eugénie von Tunzelmann, Kip, and Oliver James.

ware called Mathematica, and then sent the equations and Mathematica code to Oliver. He devoured them, converted them into sophisticated computer code that could generate the ultra-high-quality IMAX images needed for *Interstellar,* and then passed them on to Eugénie and her team. It was a joy working with them.

And the end product, the visualizations in *Interstellar,* are amazing! And scientifically accurate.

You cannot imagine how ecstatic I was when Oliver sent me his initial film clips. For the first time ever—and before any other scientist—I saw in ultrahigh definition what a fast-spinning black hole looks like. What it does, visually, to its environment.

Matthew McConaughey, Anne Hathaway, Michael Caine, Jessica Chastain

On July 18, two weeks before filming was to begin, I received an e-mail from Matthew McConaughey, who plays Cooper: "per Interstellar," he wrote, "I'd like to ask you some questions and . . . If you are around L.A. area, in person is preferable. Lemme know please, thanks, in process, mcConaughey."

We met six days later, in a suite at L'Hermitage, a boutique hotel in Beverly Hills. He was ensconced there, struggling to wrap his head around the role of Cooper and the science of *Interstellar.*

When I arrived, he opened the door dressed in shorts and a tank top, barefooted and thin from having just filmed *Dallas Buyers' Club* (for which he later won the Oscar for best actor). He asked if he could call me "Kip"; I said of course and asked what I should call him. "Anything but Matt; I hate Matt." "Matthew." "McConaughey." "Hey you." "Whatever you like." I chose "McConaughey" as it trips off the tongue so nicely, and there are too many Matthews in my life.

McConaughey had removed all the furniture from the suite's huge living/dining room, except an L-shaped couch and a coffee table. Strewn over the floor and table were 12-by-18-inch sheets of paper, each covered with notes dealing with a particular topic, written in random directions, squiwampus. We sat on the couch. He would pick up a sheet, browse it, and ask a question. The question was usually deep, and triggered a long discussion during which he would write notes on the sheet.

Often the discussion would take off in unexpected directions, with the sheet forgotten. It was one of the most interesting and enjoyable conversations I've had in a long time! We wandered from the laws of physics, especially quantum physics, to religion and mysticism, to the science of *Interstellar,* to our families and especially our children, to our philosophies of life, to how we each get inspirations, how our minds work, how we make discoveries. I left, two hours later, in a state of euphoria.

Later I told Lynda about our meeting. "Of course," she responded. She could have told me what to expect; *Interstellar* is her third film with McConaughey. I'm glad she didn't tell me. It was a joy to discover for myself.

The next e-mail, a few weeks later, was from Anne Hathaway, who plays Amelia Brand. "Hi Kip! I hope this e-mail finds you well. . . . Emma Thomas passed along your e-mail in case I had any questions. Well, the subject matter is pretty dense so I have a few! . . . would we be able to chat? Thank you very much, Annie."

We talked by phone, as our schedules couldn't be meshed for an in-person meeting. She described herself as a bit of a physics geek, and said that her character, Brand, is expected to know the physics cold—and then she launched into a series of surprisingly technical physics questions: What is the relationship of time to gravity? Why do we think there might be higher dimensions? What is the current status of research on quantum gravity? Are there any experimental tests of quantum gravity? . . . Only at the end did she let us wander off subject, to music, in fact. She played trumpet in high school; I played sax and clarinet.

During the filming of *Interstellar,* I was on set very, very little. I was not needed. But one morning Emma Thomas toured me through the *Endurance* set—a full-scale mockup of the *Endurance* spacecraft's command and navigation pod, in Stage 30 at Sony Studios.

It was tremendously impressive: 44 feet long, 26 feet wide, 16 feet high, suspended in midair; able to shift from horizontal to nearly vertical; exquisite in detail. It blew me away, and piqued my curiosity.

"Emma, why build these enormous, complex sets, when the same thing could be done with computer graphics?" "It's not clear which would be cheaper," she responded. "And computer graphics can't yet produce the compelling visual details of a real set." Wherever possible, she and Chris use real sets and real practical effects, except for things that can't actually be shot that way, like the black hole Gargantua.

On another occasion, I wrote dozens of equations and diagrams on Professor Brand's blackboards, and watched as Chris filmed in the Professor's office with Michael Caine as the Professor and Jessica Chastain as Murph.[1] I was astonished by the warm and friendly deference that Caine and Chastain showed me. Despite having no role in the filming, I was notorious as *Interstellar*'s real scientist, the guy who inspired everyone's best effort to get the science right for this blockbuster movie.

That notoriety triggered fascinating conversations with Hollywood icons: not just the Nolans, McConaughey, and Hathaway, but also Caine, Chastain, and others. A fun bonus from my creative friendship with Lynda.

Now comes the final phase of Lynda's and my *Interstellar* dream. The phase where you, the audience, have become curious about *Interstellar*'s science and seek explanations for bizarre things you saw in the movie.

The answers are here. That's why I wrote this book. Enjoy!

1 See Chapter 25.

I

FOUNDATIONS

2

Our Universe in Brief

Our universe is vast. Achingly beautiful. Remarkably simple in some ways, intricately complex in others. From our universe's great richness, we'll need only a few basic facts that I'll now lay bare.

The Big Bang

Our universe was born in a gigantic explosion 13.7 billion years ago. The explosion was given the irreverent name "the big bang" by my friend Fred Hoyle, a cosmologist who at that time (the 1940s) thought it an outrageous, fictional idea.

Fred was proved wrong. We've since seen radiation from the explosion, even in just the last week (as I write this) tentative evidence for radiation emitted in the first trillionth of a trillionth of a trillionth of a second after the explosion began![1]

We don't know what triggered the big bang, nor what, if anything, existed before it. But somehow the universe emerged as a vast sea of

1 Google "gravitational waves from the big bang" or "CMB polarization" to learn about this amazing March 2014 discovery. I give some details at the end of Chapter 16.

ultrahot gas, expanding fast in all directions like the fireball ignited by a nuclear bomb blast or by the explosion of a gas pipeline. Except that the big bang was not destructive (so far as we know). Instead, it *created* everything in our universe, or rather the seeds for everything.

I would love to write a long chapter about the big bang, but with great force of will I restrain myself. We don't need it for the rest of this book.

Galaxies

As our universe expanded, its hot gas cooled. In some regions the gas's density was a bit higher than in others, randomly. When the gas got cold enough, gravity pulled each high-density region inward on itself, giving birth to a galaxy (a huge cluster of stars and their planets and diffuse gas between the stars); see Figure 2.1. The earliest galaxy was born when the universe was a few hundred million years old.

There are roughly a trillion galaxies in the visible universe. The largest

Fig. 2.1. A rich cluster of galaxies named Abell 1689 and many other more distant galaxies, as photographed by the Hubble Space Telescope.

galaxies contain a few trillion stars and are about a million light-years across;[2] the smallest, about 10 million stars and a thousand light-years across. At the center of most every large galaxy there is a huge black hole (Chapter 5), one that weighs a million times the sun's weight or more.[3]

The Earth resides in a galaxy called the Milky Way. Most of the Milky Way's stars are in the bright band of light that stretches across Earth's sky on a clear, dark night. And almost all the pinpricks of light that we see in the sky at night, not just those in the bright band, also lie in the Milky Way.

The nearest large galaxy to our own is called Andromeda (Figure 2.2). It is 2.5 million light-years from Earth. It contains about a trillion stars and is about 100,000 light-years across. The Milky Way is a sort of twin to Andromeda, about the same in size, shape, and number of stars. If Figure 2.2 were the Milky Way, then the Earth would be where I placed the yellow diamond.

Andromeda contains a gigantic black hole, 100 million times

Fig. 2.2. The Andromeda galaxy.

2 A light-year is the distance light travels in one year: about a hundred trillion kilometers.

3 In more technical language, its mass is a million times that of the Sun's or more, which means its gravitational pull, when you are at some fixed distance away from it, is the same as a million Suns'. In this book I use "mass" and "weight" to mean the same thing.

heavier than the Sun and as big across as the Earth's orbit (the same weight and size as *Interstellar*'s Gargantua; Chapter 6). It resides in the middle of the central bright sphere in Figure 2.2.

Solar System

Stars are large, hot balls of gas, usually kept hot by burning nuclear fuel in their cores. The Sun is a fairly typical star. It is 1.4 million kilometers across, about a hundred times larger than the Earth. Its surface has flares and hot spots and cooler spots, and is fascinating to explore through a telescope (Figure 2.3).

Eight planets, including the Earth, travel around the Sun in elliptical orbits, along with many dwarf planets (of which Pluto is the most famous) and many comets, and smaller, rocky bodies called asteroids

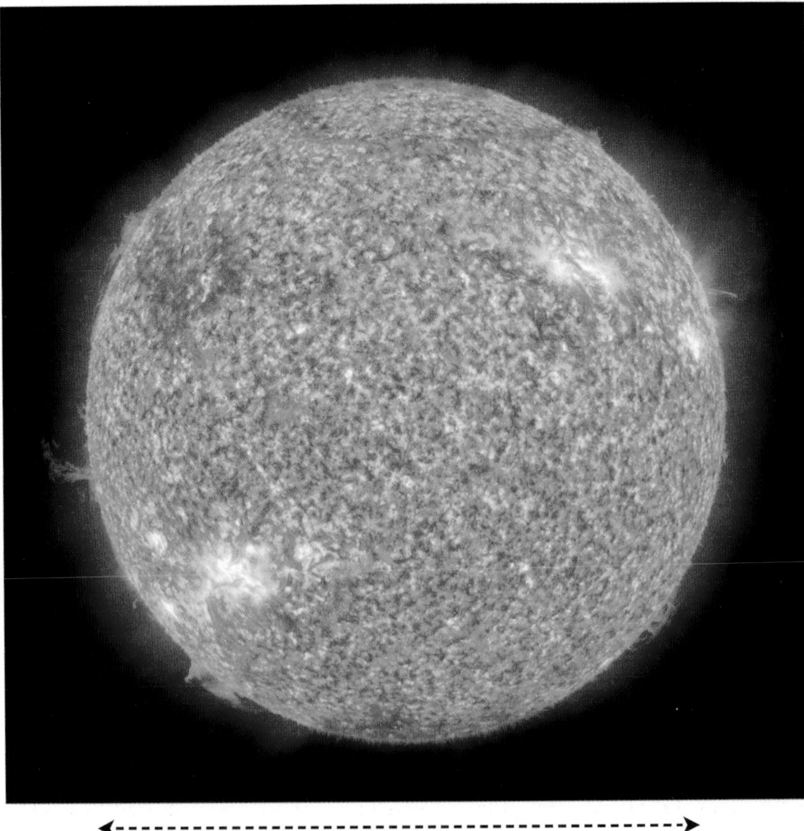

Fig 2.3. The Sun as photographed by NASA's Solar Dynamics Observatory.

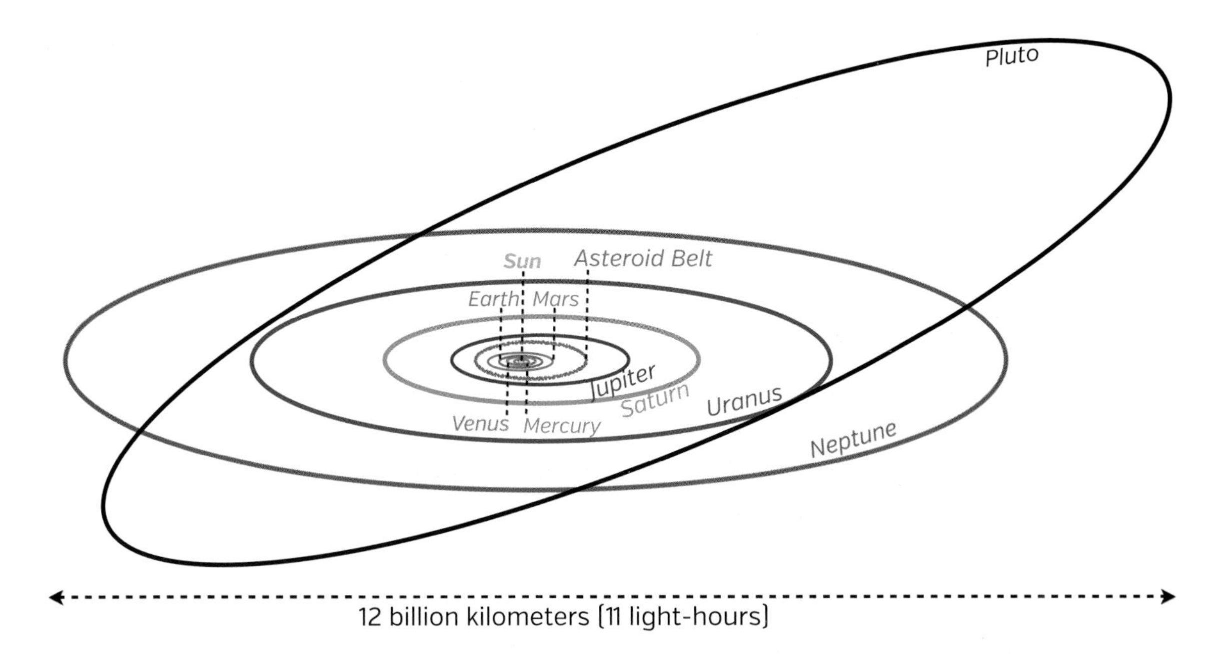

Fig. 2.4. The orbits of the Sun's planets and Pluto, and a region containing many asteroids.

and meteoroids (Figure 2.4). Earth is the third planet from the Sun. Saturn, with its gorgeous rings, is the sixth planet out and plays a role in *Interstellar* (Chapter 15).

The solar system is a thousand times bigger than the Sun itself; light needs eleven hours to travel across it.

The distance to the nearest star other than the Sun, Proxima Centauri, is 4.24 light-years, 2500 times farther than the distance across the solar system! In Chapter 13, I discuss the awful implications for interstellar travel.

Stellar Death: White Dwarfs, Neutron Stars, and Black Holes

The Sun and Earth are about 4.5 billion years old, about a third the age of the universe. After another 6.5 billion years or so, the Sun will exhaust the nuclear fuel in its core, the fuel that keeps it hot. The Sun then will shift to burning fuel in a shell around its core, and its surface will expand to engulf and fry the Earth. With the shell's fuel spent and the Earth fried, the Sun will shrink to become a white dwarf star, about the size of the Earth but with density a million times higher. The white dwarf will gradually cool, over tens of billions of years, to become a dense, dark cinder.

Stars much heavier than the Sun burn their fuel much more quickly, and then collapse to form a neutron star or a black hole.

Neutron stars have masses about one to three times that of the Sun, circumferences of 75 to 100 kilometers (about the size of Chicago), and densities the same as the nucleus of an atom: a hundred trillion times more dense than rock and the Earth. Indeed, neutron stars are made of almost pure nuclear matter: atomic nuclei packed side by side.

Black holes (Chapter 5), by contrast, are made fully and solely from warped space and warped time (I'll explain this weird claim in Chapter 4). They contain no matter whatsoever, but they have surfaces, called "event horizons," or just "horizons," through which nothing can escape, not even light. That's why they are black. A black hole's circumference is proportional to its mass: the heavier it is, the bigger it is.

A black hole with about the same mass as a typical neutron star or white dwarf (say 1.2 times as heavy as the Sun) has a circumference of about 22 kilometers: a fourth that of the neutron star and a thousandth that of the white dwarf. See Figure 2.5.

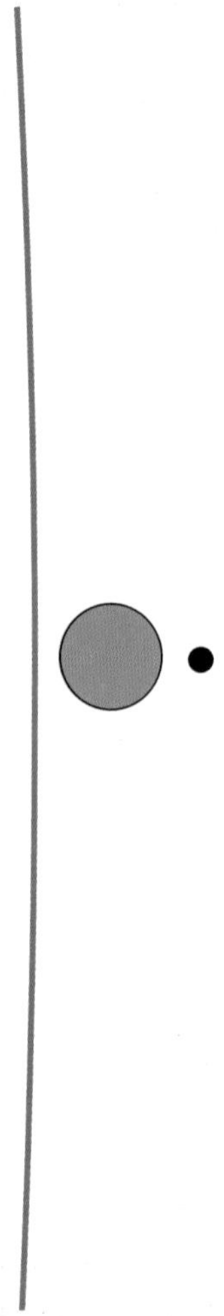

Fig. 2.5. A white dwarf (*left*), neutron star (*middle*), and black hole (*right*) that all weigh as much as 1.2 Suns. For the white dwarf I show only a tiny segment of its surface.

Since stars are generally no heavier than about 100 Suns, the black holes to which they give birth are also no heavier than 100 Suns. The giant black holes in the cores of galaxies, a million to 20 billion times heavier than the Sun, therefore, cannot have been born in the death of a star. They must have formed in some other way, perhaps by the agglomeration of many smaller black holes; perhaps by the collapse of massive clouds of gas.

Magnetic, Electric, and Gravitational Fields

Because magnetic force lines play a big role in our universe and are important for *Interstellar*, let's discuss them, too, before diving into *Interstellar*'s science.

As a student in science class, you may have met magnetic force lines in a beautiful little experiment. Do you remember taking a sheet of paper, placing a bar magnet under it, and sprinkling iron filings (elongated flakes of iron) on top of the paper? The iron filings make

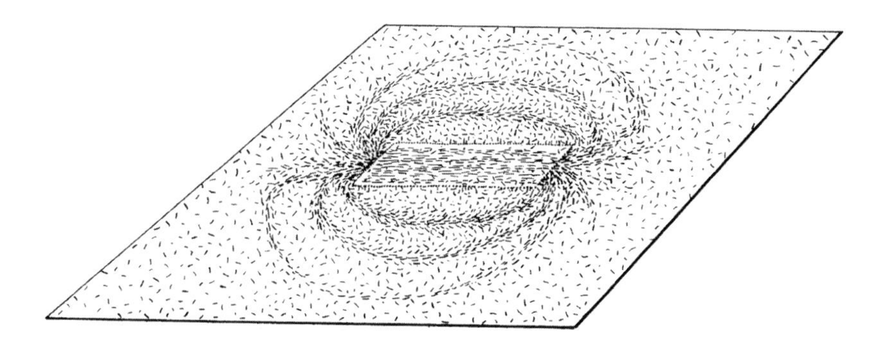

Fig. 2.6. Magnetic force lines from a bar magnet, made visible by iron filings sprinkled on a sheet of paper. *[Drawing by Matt Zimet based on a sketch by me; from my book* Black Holes & Time Warps: Einstein's Outrageous Legacy.*]*

the pattern shown in Figure 2.6. They orient themselves along magnetic force lines that otherwise are invisible. The force lines depart from one of the magnet's poles, swing around the magnet, and descend into the other pole. The magnetic *field* is the collection of all the magnetic force lines.

When you try to push two magnets together with their north poles facing each other, their force lines repel each other. You see nothing between the magnets, but you feel the magnetic field's repulsive force. This can be used for magnetic levitation, suspending a magnetized object—even a railroad train (Figure 2.7)—in midair.

The Earth also has two magnetic poles, north and south. Magnetic

Fig. 2.7. The world's first commercial magnetically levitated train, in Shanghai, China.

Fig. 2.8. The Earth's magnetic force lines.

force lines depart from the south magnetic pole, swing around the Earth, and descend into the north magnetic pole (Figure 2.8). These force lines grab a compass needle, just as they grab iron filings, and drive the needle to point as nearly along the force lines as possible. That's how a compass works.

The Earth's magnetic force lines are made visible by the Aurora Borealis (the Northern Lights; Figure 2.9). Protons flying outward

Fig. 2.9. The Aurora Borealis in the sky over Hammerfest, Norway.

Fig. 2.10. Artist's conception of a neutron star with its donut-shaped magnetic field and its jets.

from the Sun are caught by the force lines and travel along them into the Earth's atmosphere. There the protons collide with oxygen and nitrogen molecules, making the oxygen and nitrogen fluoresce. That fluorescent light is the Aurora.

Neutron stars have very strong magnetic fields, whose force lines are donut-shaped, like the Earth's. Fast-moving particles trapped in a neutron star's magnetic field light up the force lines, producing the blue rings in Figure 2.10. Some of the particles are liberated and stream out the field's poles, producing the two violet jets in the figure. These jets consist of all types of radiation: gamma rays, X-rays; ultraviolet, visual, infrared, and radio waves. As the star spins, its luminous jets sweep around the sky above the neutron star, like a searchlight. Every time a jet sweeps over the Earth, astronomers see a pulse of radiation, so astronomers have named these objects "pulsars."

The universe contains other kinds of fields (collections of force lines) in addition to magnetic fields. One example is electric fields (collections of electric force lines that, for example, drive electric current to flow through wires). Another example is gravitational fields (collections of gravitational force lines that, for example, pull us to the Earth's surface).

The Earth's gravitational force lines point radially into the Earth and they pull objects toward the Earth along themselves. The strength of the gravitational pull is proportional to the density of the force lines (the number of lines passing through a fixed area). As they reach inward, the force lines pass through spheres of ever-decreasing area (dotted red spheres in Figure 2.11), so the lines' density must go up inversely with the sphere's area, which means the Earth's gravity grows as you travel toward it, as 1/(the red spheres' area). Since each sphere's area is proportional to the square of its distance r from the Earth's center, the strength of the Earth's gravitational pull grows as $1/r^2$. This is Newton's inverse square law for gravity—an example of the fundamental laws of physics that are Professor Brand's passion in *Interstellar* and our next foundation for *Interstellar*'s science.

Fig. 2.11. The Earth's gravitational force lines.

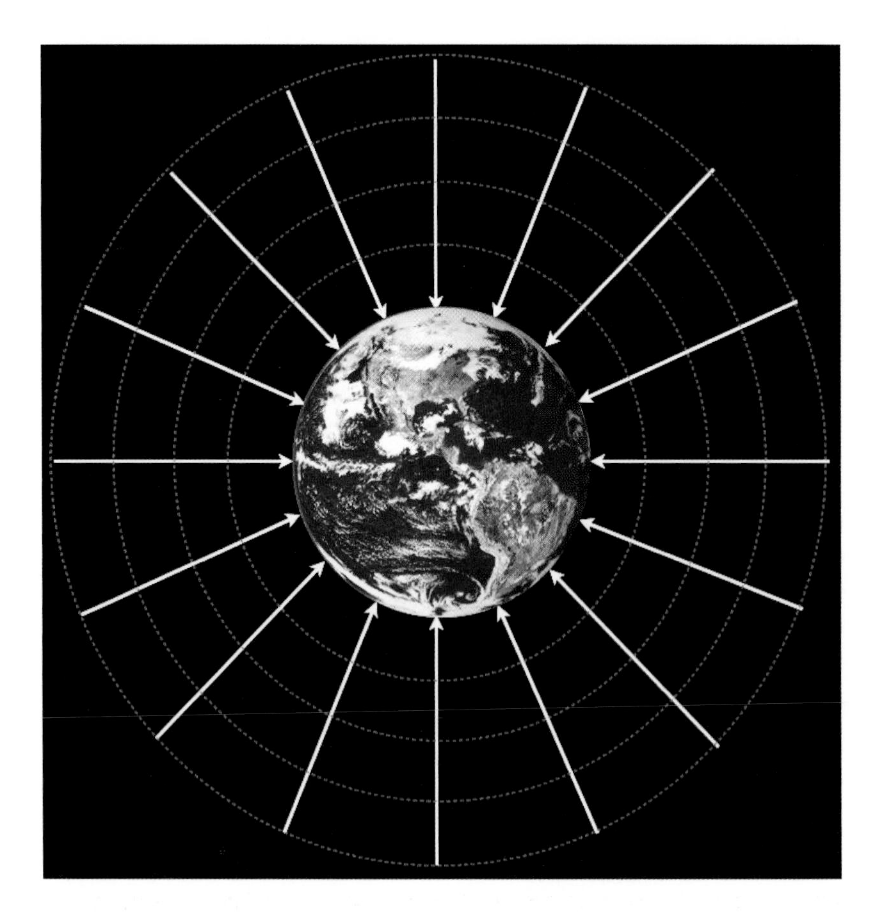

3

The Laws That Control the Universe

Mapping the World and Deciphering the Laws of Physics

Physicists have struggled from the seventeenth century onward to discover the physical laws that shape and control our universe. This has been like European explorers struggling to discover the Earth's geography (Figure 3.1).

By 1506 Eurasia was coming into focus and there were glimmers of South America. By 1570 the Americas were coming into focus, but there was no sign of Australia. By 1744 Australia was coming into focus, but Antarctica was terra incognita.

Similarly (Figure 3.2), by 1690 the *Newtonian laws of physics* had come into focus. With concepts such as force, mass, and acceleration and equations that link them, such as $F = ma$, the Newtonian laws accurately describe the motion of the Moon around the Earth and the Earth around the Sun, the flight of an airplane, the construction of a bridge, and collisions of a child's marbles. In Chapter 2 we briefly met an example of a Newtonian law, the inverse square law for gravity.

By 1915 Einstein and others had found strong evidence that the Newtonian laws fail in the realm of the very fast (objects that move at nearly the speed of light), the realm of the very large (our universe as a whole), and

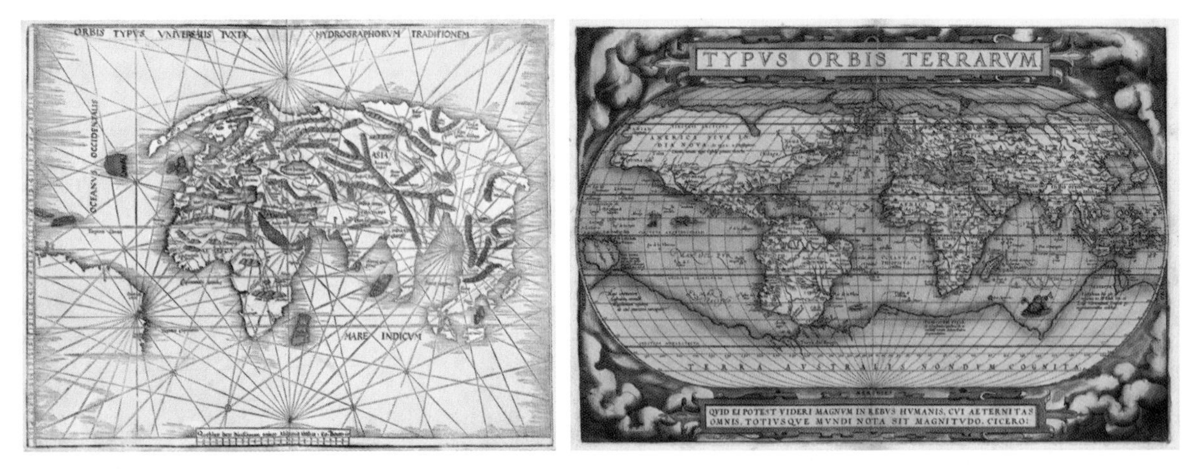

1506–Martin Waldseemuller

1570–Abraham Ortelius

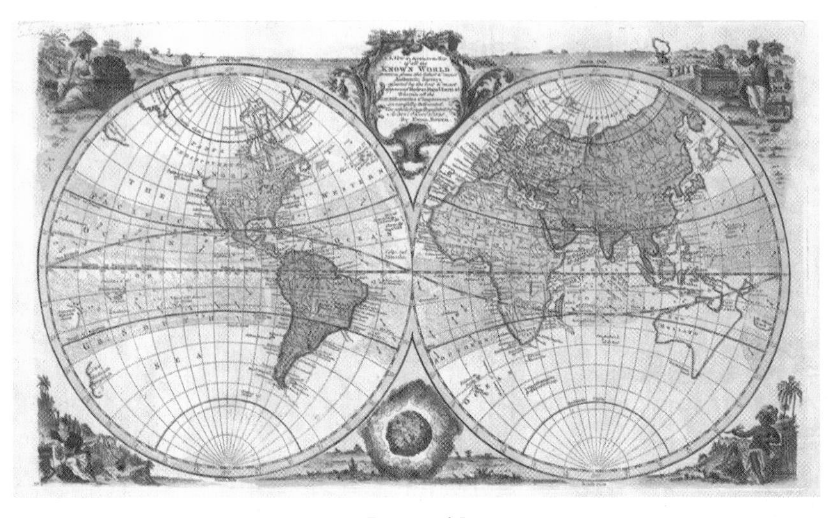

Fig. 3.1. World maps from 1506 to 1744.

1744–Emanuel Bowen

the realm of intense gravity (for example, black holes). To remedy these failures Einstein gave us his revolutionary *relativistic laws of physics* (Figure 3.2). Using the concepts of warped time and warped space (which I describe in the next chapter), the relativistic laws predicted and explained the expansion of the universe, black holes, neutron stars, and wormholes.

By 1924 it was crystal clear that the Newtonian laws also fail in the realm of the very small (molecules, atoms, and fundamental particles). To deal with this Niels Bohr, Werner Heisenberg, Erwin Schrödinger, and others gave us the *quantum laws of physics* (Figure 3.2). Using the concepts that everything fluctuates randomly at least a little bit (which I describe in Chapter 26), and that these fluctuations can produce new particles and radiation where before there were none, the

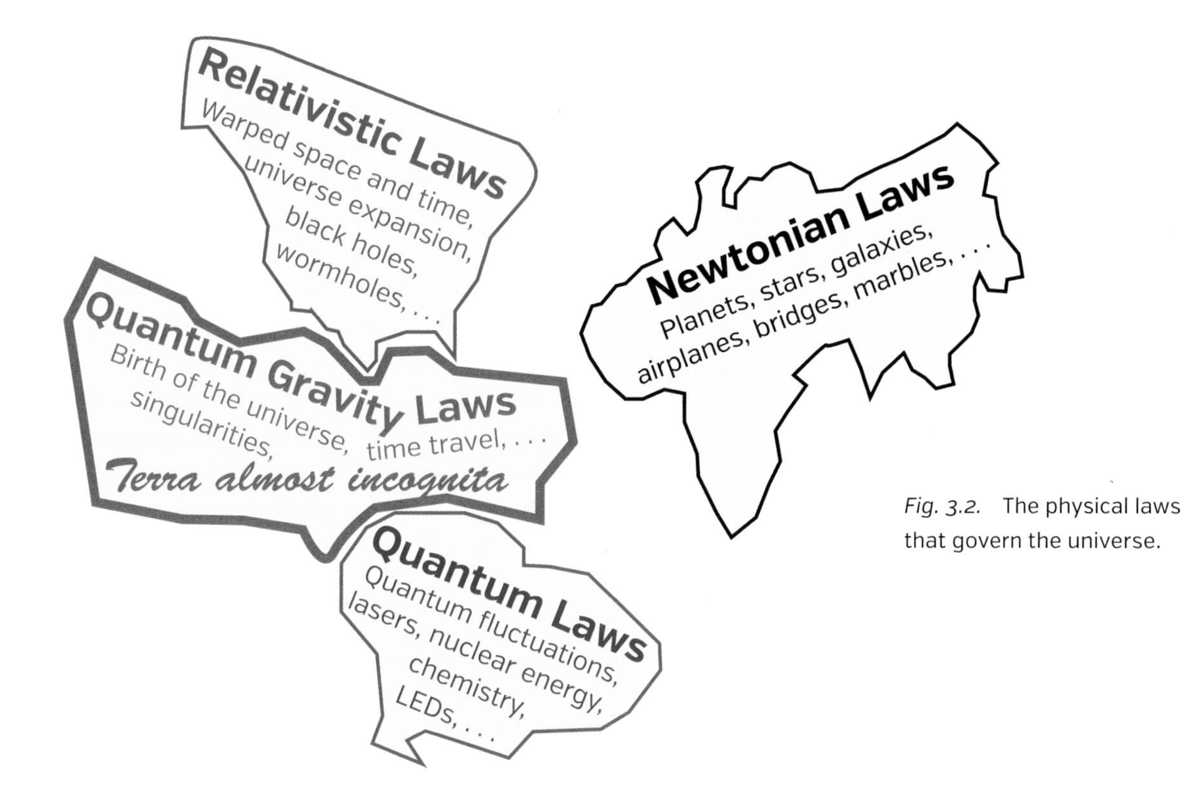

Fig. 3.2. The physical laws that govern the universe.

quantum laws have brought us lasers, nuclear energy, light-emitting diodes, and a deep understanding of chemistry.

By 1957 it became evident that the relativistic laws and the quantum laws are fundamentally incompatible. They predict different things, incompatible things, in realms where gravity is intense *and* quantum fluctuations are strong.[1] These realms include the big bang birth of our universe (Chapter 2), the cores of black holes like Gargantua (Chapters 26 and 28), and backward time travel (Chapter 30). In these realms a "fiery marriage"[2] of the incompatible relativistic and quantum laws gives rise to new *laws of quantum gravity* (Figure 3.2).

1 In these realms, for example, the energy of light has huge quantum fluctuations. They are so huge that they warp space and time enormously and randomly. The fluctuating warpage is beyond the scope of Einstein's relativistic laws, and the warpage's influence on the light is beyond the scope of the light's quantum laws.

2 The phrase "fiery marriage" was coined by my mentor John Wheeler, who was superb at naming things. John also coined the words "black hole" and "wormhole" and the phrase "a black hole has no hair"; Chapters 14 and 5. He once described to me lying in a warm bath for hours on end, letting his mind soar in a search for just the right word or phrase.

We do not yet know the laws of quantum gravity, but we have some compelling insights, including superstring theory (Chapter 21), thanks to enormous effort by the world's greatest twenty-first-century physicists. Despite those insights, quantum gravity remains terra almost incognita (an almost unknown land). This leaves much elbow room for exciting science fiction, elbow room that Christopher Nolan exploits with great finesse in *Interstellar*; see Chapters 28–31.

Truth, Educated Guesses, and Speculations

The science of *Interstellar* lies in all four domains: Newtonian, relativistic, quantum, and quantum gravity. Correspondingly, some of the science is known to be true, some is an educated guess, and some is speculation.

To be *true*, the science must be based on well-established physical laws (Newtonian, relativistic, or quantum), and it must have enough basis in observation that we are confident of how to apply the well-established laws.

In precisely this sense, neutron stars and their magnetic fields, as described in Chapter 2, are true. Why? First, neutron stars are firmly predicted to exist by the quantum and relativistic laws. Second, astronomers have studied in enormous detail the pulsar radiation from neutron stars (pulses of light, X-rays, and radio waves described in Chapter 2). These pulsar observations are beautifully and accurately explained by the quantum and relativistic laws, if the pulsar is a spinning neutron star; and no other explanation has ever been found. Third, neutron stars are firmly predicted to form in astronomical explosions called supernovae, and pulsars are seen at the centers of big, expanding gas clouds, the remnants of old supernovae. Thus, we astrophysicists have no doubt; neutron stars really do exist and they really do produce the observed pulsar radiation.

Another example of a truth is the black hole Gargantua and the bending of light rays by which it distorts images of stars (Figure 3.3). Physicists call this distortion "gravitational lensing" because it is similar to the distortion of a picture by a curved lens or mirror, as in an amusement park's fun house, for example.

Einstein's relativistic laws predict, unequivocally, all the properties of black holes from their surfaces outward, including their gravitational lensing.[3] Astronomers have firm observational evidence that black holes exist in our universe, including gigantic black holes like Gargantua. Astronomers have seen gravitational lensing by other objects (for example, Figure 24.3), though not yet by black holes, and the observed lensing is in precise accord with the predictions of Einstein's relativistic laws. This is enough for me. Gargantua's gravitational lensing, as simulated by Paul Franklin's Double Negative team using relativity equations I gave to them, is true. This is what it really would look like.

Fig. 3.3. The stars in Gargantua's galaxy, as seen around Gargantua's shadow. Gargantua bends the light rays coming from each star, thereby distorting enormously the appearance of its galaxy: "gravitationally lensing" the galaxy. *[From a simulation for this book by the Double Negative visual-effects team.]*

By contrast, the blight that endangers human life on Earth in *Interstellar* (Figure 3.4 and Chapter 11) is an educated guess in one sense, and a speculation in another. Let me explain.

Throughout recorded history, the crops that humans grow have been plagued by occasional blights (rapidly spreading diseases caused by

Fig. 3.4. Burning blighted corn. *[From* Interstellar, *used courtesy of Warner Bros. Entertainment Inc.]*

3 Chapters 5, 6, and 8.

microbes). The biology that underlies these blights is based on chemistry, which in turn is based on the quantum laws. Scientists do not yet know how to deduce, from the quantum laws, *all* of the relevant chemistry (but they can deduce *much* of it); and they do not yet know how to deduce from chemistry all of the relevant biology. Nevertheless, from observations and experiments, biologists have learned much about blights. The blights encountered by humans thus far have not jumped from infecting one type of plant to another with such speed as to endanger human life. But nothing we know guarantees this can't happen. That such a blight is possible is an *educated guess*. That it might someday occur is a *speculation* that most biologists regard as very unlikely.

The gravitational anomalies that occur in *Interstellar* (Chapters 24 and 25), for example, the coin Cooper tosses that suddenly plunges to the floor, are *speculations*. So is harnessing the anomalies to lift colonies off Earth (Chapter 31).

Although experimental physicists when measuring gravity have searched hard for anomalies—behaviors that cannot be explained by the Newtonian or relativistic laws—no convincing gravitational anomalies have ever been seen on Earth.

However, it seems likely from the quest to understand quantum gravity that our universe is a membrane (physicists call it a "brane") residing in a higher-dimensional "hyperspace" to which physicists give the name "bulk"; see Figure 3.5 and Chapters 4 and 21. When physicists carry Einstein's relativistic laws into this bulk, as Professor Brand does on the blackboard in his office (Figure 3.6), they discover the possibility of gravitational anomalies—anomalies triggered by physical fields that reside in the bulk.

We are far from sure that the bulk really exists. And it is only an educated guess that, if the bulk does exist, Einstein's laws reign there. And we have no idea whether the bulk, if it exists, contains fields that can

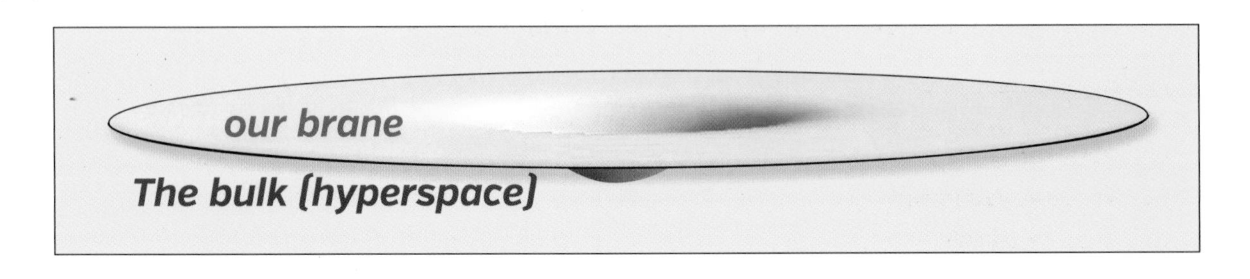

Fig. 3.5. Our universe, in the vicinity of the Sun, depicted as a two-dimensional surface or brane, residing in a three-dimensional bulk. In reality, our brane has three space dimensions and the bulk has four. This figure is explained further in Chapter 4; see especially Figure 4.4.

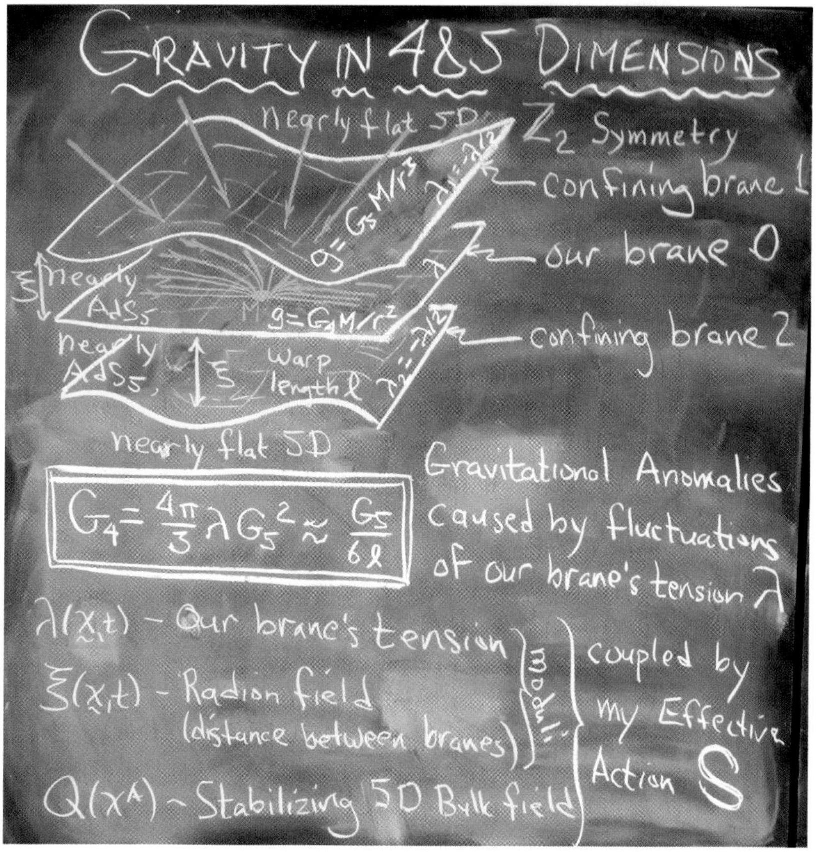

Fig. 3.6. Relativity equations on Professor Brand's blackboard, describing possible foundations for gravitational anomalies. For details see Chapter 25.

generate gravitational anomalies, and if so, whether those anomalies can be harnessed. The anomalies and their harnessing are a rather extreme speculation. But they are a speculation based on science that I and some of my physicist friends are happy to entertain—at least late at night over beer. So they fall within the guidelines I advocated for *Interstellar:* "Speculations . . . will spring from real science, from ideas that at least some 'respectable' scientists regard as possible" (Chapter 1).

Throughout this book, when discussing the science of *Interstellar,* I explain the status of that science—truth, educated guess, or speculation—and I label it so at the beginning of a chapter or section with a symbol:

Ⓣ for truth
(EG) for educated guess
△S for speculation

Of course, the status of an idea—truth, educated guess, or speculation—can change; and you'll meet such changes occasionally in the movie and in this book. For Cooper, the bulk is an educated guess that becomes a truth when he goes there in the tesseract (Chapter 29); and the laws of quantum gravity are a speculation until TARS extracts them from inside a black hole so for Cooper and Murph they become truth (Chapters 28 and 30).

For nineteenth-century physicists, Newton's inverse square law for gravity was an absolute truth. But around 1890 it was revolutionarily upended by a tiny observed anomaly in the orbit of Mercury around the Sun (Chapter 24). Newton's law is very nearly correct in our solar system, but not quite. This anomaly helped pave the way for Einstein's twentieth-century relativistic laws, which—in the realm of strong gravity—began as speculation, became an educated guess when observational data started rolling in, and by 1980, with ever-improving observations, evolved into truth (Chapter 4).

Revolutions that upend established scientific truth are exceedingly rare. But when they happen, they can have profound effects on science and technology.

Can you identify in your own life speculations that became educated guesses and then truth? Have you ever seen your established truths upended, with a resulting revolution in your life?

4

Warped Time and Space, and Tidal Gravity

Ⓣ

Einstein's Law of Time Warps

Einstein struggled to understand gravity on and off from 1907 onward. Finally in 1912 he had a brilliant inspiration. Time, he realized, must be warped by the masses of heavy bodies such as the Earth or a black hole, and that warping is responsible for gravity. He embodied this insight in what I like to call "Einstein's law of time warps," a precise mathematical formula[1] that I describe qualitatively this way: *Everything likes to live where it will age the most slowly, and gravity pulls it there.*

The greater the slowing of time, the stronger gravity's pull. On Earth, where time is slowed by only a few microseconds per day, gravity's pull is modest. On the surface of a neutron star, where time is slowed by a few hours per day, gravity's pull is enormous. At the surface of a black hole, time is slowed to a halt, whence gravity's pull is so humungous that nothing can escape, not even light.

This slowing of time near a black hole plays a major role in *Interstellar.* Cooper despairs of ever seeing his daughter Murph again, when

1 See *Some Technical Notes* at the end of this book.

his travel near Gargantua causes him to age only a few hours while Murph, on Earth, is aging eight decades.

Human technology was too puny to test Einstein's law until nearly half a century after he formulated it. The first good test came in 1959 when Bob Pound and Glen Rebca used a new technique called the Mössbauer effect to compare the rate of flow of time in the basement of a 73-foot tower at Harvard University with time in the tower's penthouse. Their experiment was exquisitely accurate: good enough to detect differences of 0.0000000000016 seconds (1.6 trillionths of a second) in one day. Remarkably, they found a difference 130 times larger than this accuracy and in excellent agreement with Einstein's law: Time flows more slowly in the basement than in the penthouse by 210-trillionths of a second each day.

The accuracy improved in 1976, when Robert Vessot of Harvard flew an atomic clock on a NASA rocket to a 10,000-kilometer height, and used radio signals to compare its ticking rate with clocks on the ground (Figure 4.1). Vessot found that time on the ground flows more slowly than at a height of 10,000 kilometers by about 30 microseconds (0.00003 seconds) in one day, and his measurement agreed with Einstein's law of time warps to within his experimental accuracy. That accuracy (the uncertainty in Vessot's measurement) was seven parts in a hundred thousand: 0.00007 of 30 microseconds in a day.

Fig. 4.1. Atomic clocks measure slowing of time on Earth. *[Reproduced from* Was Einstein Right? Putting General Relativity to the Test, *by Clifford M. Will (Basic Books, 1993).]*

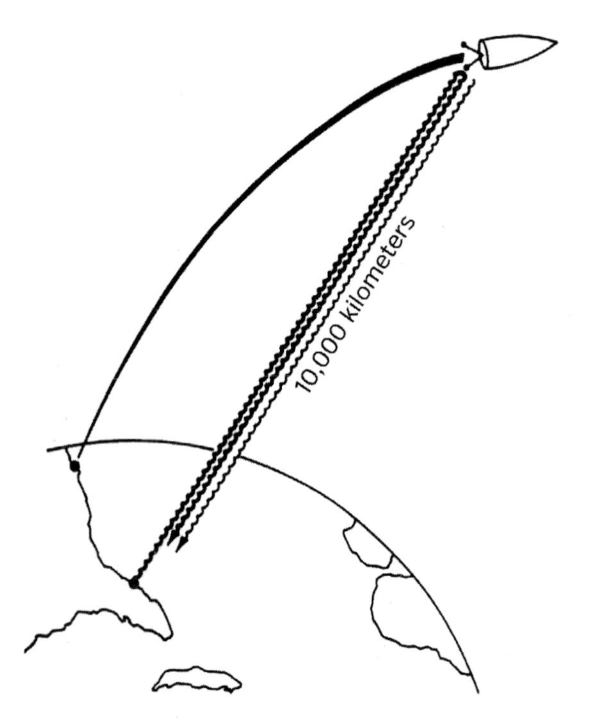

The global positioning system (GPS), by which our smart phones can tell us where we are to 10 meters' accuracy, relies on radio signals from a set of 27 satellites at a height of 20,000 kilometers (Figure 4.2). Typically only four to twelve satellites can be seen at once from any location on Earth. Each radio signal from a viewable satellite tells the smart phone where the satellite is located and the time the signal was transmitted. The smart phone measures the signal's arrival time and compares it with its transmission time to learn how far the signal traveled—the distance between satellite and phone. Knowing the locations and distances to several satellites, the smart phone can triangulate to learn its own location.

This scheme would fail if the signal transmis-

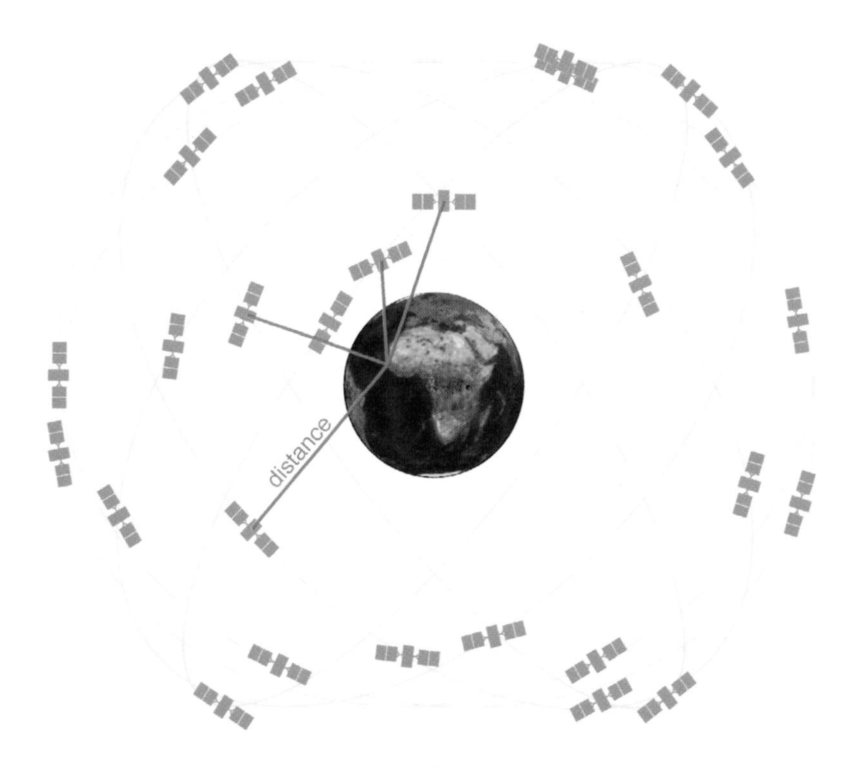

Fig. 4.2. The global positioning system.

sion times were the true times measured on the satellite. Time at a 20,000-kilometer height flows more rapidly than on Earth by forty microseconds each day, and the satellites must correct for this. They measure time with their own clocks, then slow that time down to the rate of time flow on Earth before transmitting it to our phones.

Einstein was a genius. Perhaps the greatest scientist ever. This is one of many examples where his insights about the laws of physics could not be tested in his own day. It required a half century for technology to improve enough for a test with high precision, and another half century until the phenomena he described became part of everyday life. Among other examples are the laser, nuclear energy, and quantum cryptography.

The Warping of Space: The Bulk and Our Brane

In 1912 Einstein realized that if time can be warped by massive bodies, then space must also be warped. But despite the most intense mental struggle of his life, the full details of space warps long eluded him.

From 1912 to late 1915 he struggled. Finally in November 1915, in a great Eureka moment, he formulated his "field equation of general relativity," which encapsulated all his relativistic laws including space warps.

Again, human technology was too puny for high-precision tests.[2] This time the needed improvements took sixty years, culminating in several key experiments. The one I liked best was led by Robert Reasenberg and Irwin Shapiro of Harvard. In 1976–77 they transmitted radio signals to two spacecraft in orbit around Mars. The spacecraft, called *Viking 1* and *Viking 2*, amplified the signals and sent them back to Earth, where their round-trip travel time was measured. As the Earth and Mars moved around the Sun in their orbits, the radio signals traversed paths that were changing. At first, the paths were far from the Sun, then they passed near the Sun, and then far again, as shown in the bottom half of Figure 4.3.

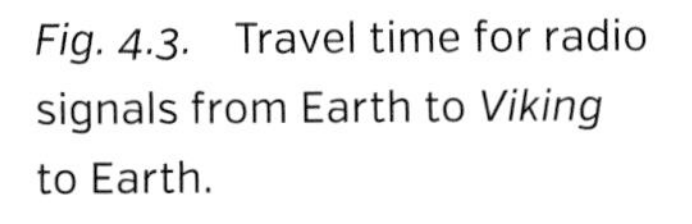

Fig. 4.3. Travel time for radio signals from Earth to *Viking* to Earth.

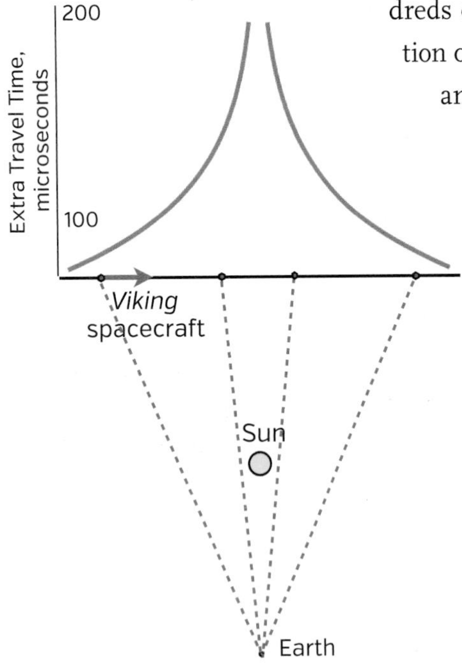

If space were flat, the round-trip travel time would have changed gradually and steadily. It did not. When the radio waves passed near the Sun, their travel time was longer than expected, longer by hundreds of microseconds. The extra travel time is shown, as a function of the spacecraft's location at the top of Figure 4.3; it went up and then back down. Now, one of Einstein's relativistic laws says that radio waves and light travel at an absolutely constant, unchanging speed.[3] Therefore, the distance from Earth to the spacecraft had to be longer than expected when passing near the Sun, longer by hundreds of microseconds times the speed of light: about 50 kilometers.

This greater length would be impossible if space were flat, like a sheet of paper. It is produced by the Sun's space warp. From the extra time delay and how it changed as the spacecraft moved relative to Earth, Reasenberg and Shapiro inferred the shape of the space warp. More precisely, they inferred the shape of the two-dimensional surface formed by the paths of the

2 But see the first section of Chapter 24.

3 Unchanging after well-understood corrections for a bit of slowdown due to interaction with electrons in interplanetary space—so-called "plasma corrections."

Viking radio signals. That surface was very nearly the Sun's equatorial plane, so I describe it that way here.

The shape that the team measured, for the Sun's equatorial plane, is shown in Figure 4.4 with the magnitude of the warping exaggerated. The measured shape was precisely what Einstein's relativistic laws predict—precise to within the experimental error, which was 0.001 of the actual warping, that is, a part in a thousand. Around a neutron star, the space warp is far greater. Around a black hole, it is enormously greater.

Now, the Sun's equatorial plane divides space into two identical halves, that above the plane and that below. Nonetheless, Figure 4.4 shows the equatorial plane as warped like the surface of a bowl. It bends downward inside and near the Sun, so that diameters of circles around the Sun, when multiplied by π (3.14159 . . .), are larger than circumferences—larger, in the case of the Sun, by roughly 100 kilometers. That's not much, but it was easily measured by the spacecraft, with a precision of a part in a thousand.

How can space "bend down"? Inside *what* does it bend? It bends inside a higher-dimensional hyperspace, called "the bulk," that is not part of our universe!

Let's make that more precise. In Figure 4.4 the Sun's equatorial plane is a two-dimensional surface that bends downward in a three-dimensional bulk. This motivates the way we physicists think about our full universe. Our universe has three space dimensions (east-west, north-south, up-down), and we think of it as a three-dimensional membrane or *brane* for short that is warped in a higher-dimensional *bulk*. How many dimensions does the bulk have? I discuss this carefully in Chapter 21, but for the purposes of *Interstellar*, the bulk has just one extra space dimension: four space dimensions in all.

Now, it's very hard for humans to visualize our three-dimensional

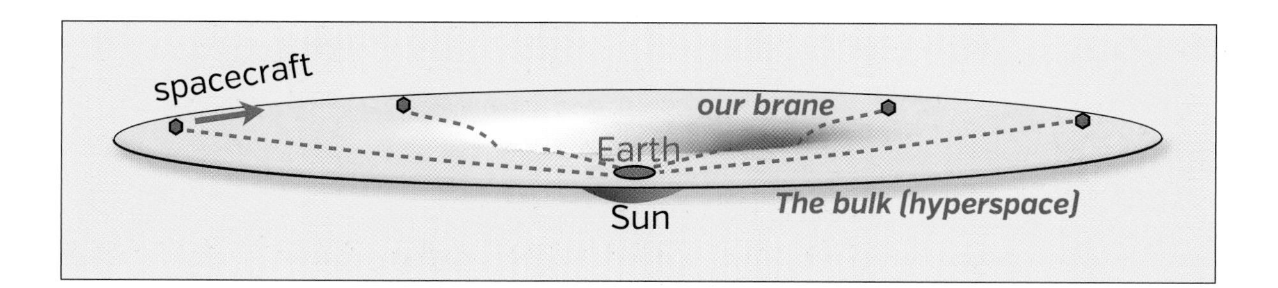

Fig. 4.4. Paths of *Viking* radio signals through the Sun's warped equatorial plane.

universe, our full brane, living and bending in a four-dimensional bulk. So throughout this book I draw pictures of our brane and bulk with one dimension removed, as I did in Figure 4.4.

In *Interstellar,* the characters often refer to *five* dimensions. Three are the space dimensions of our own universe or brane (east-west, north-south, up-down). The fourth is time, and the fifth is the bulk's extra space dimension.

Does the bulk really exist? Is there truly a fifth dimension, and maybe even more, that humans have never experienced? Very likely yes. We'll explore this in Chapter 21.

The warping of space (warping of our brane) plays a huge role in *Interstellar.* For example, it is crucial to the very existence of the wormhole connecting our solar system to the far reaches of the universe, where Gargantua lives. And it distorts the sky around the wormhole and around the black hole Gargantua; this is the gravitational lensing we met in Figure 3.3.

Figure 4.5 is an extreme example of space warps. It is a fanciful drawing by my artist friend Lia Halloran, depicting a hypothetical

Fig. 4.5. Black holes and wormholes extending out of our brane into and through the bulk. One space dimension is removed from both our brane and the bulk. *[Drawing by the artist Lia Halloran.]*

region of our universe that contains large numbers of wormholes (Chapter 14) and black holes (Chapter 5) extending outward from our brane into and through the bulk. The black holes terminate in sharp points called "singularities." The wormholes connect one region of our brane to another. As usual, I suppress one of our brane's three dimensions, so the brane looks like a two-dimensional surface.

Tidal Gravity

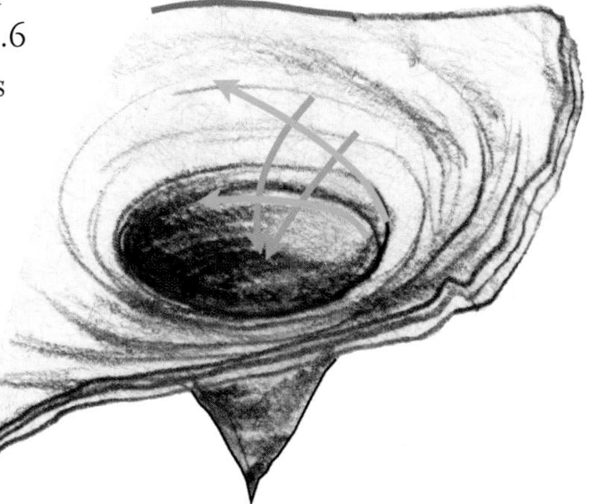

Fig. 4.6. Four paths for planetary motion in the vicinity of a black hole. The picture of the hole is extracted from Lia Halloran's drawing, Figure 4.5.

Einstein's relativistic laws dictate that planets, stars, and unpowered spacecraft near a black hole move along the straightest paths permitted by the hole's warped space and time. Figure 4.6 shows examples of four such paths. The two purple paths headed into the black hole begin parallel to each other. As each path tries to remain straight, the two paths get driven toward each other. The warping of space and time drives them together. The green paths, traveling circumferentially around the hole, also begin parallel. But in this case, the warping drives them apart.

Several years ago, my students and I discovered a new point of view about these planetary paths. In Einstein's relativity theory there is a mathematical quantity called the Riemann tensor. It describes the details of the warping of space and time. We found, hidden in the mathematics of this Riemann tensor, lines of force that squeeze some planetary paths together and stretch others apart. "Tendex lines," my student David Nichols dubbed them, from the Latin word *tendere* meaning "to stretch."

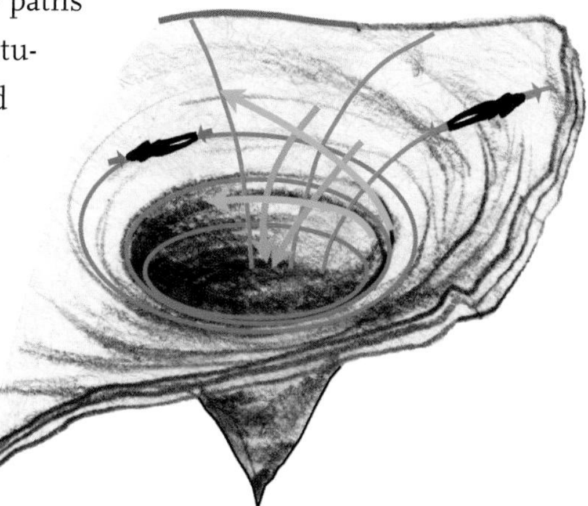

Fig. 4.7. Tendex lines around a black hole. The picture of the hole is extracted from Lia Halloran's drawing, Figure 4.5.

Figure 4.7 shows several of these tendex lines around the black hole of Figure 4.6. The green paths begin, on their right ends, parallel to each other, and then the red tendex lines stretch them apart. I draw a woman lying on a red tendex line. It stretches her, too; she feels a stretching force between her head and her feet, exerted by the red tendex line.

The purple paths begin, at their top ends, running parallel to each other. They are then squeezed together by the blue tendex lines, and the woman whose body lies along a blue tendex line is also squeezed.

This stretching and squeezing is just a different way of thinking about the influence of the warping of space and time. From one viewpoint, the paths are stretched apart or squeezed together due to the planetary paths moving along the straightest routes possible in the warped space and time. From another viewpoint it is the tendex lines that do the stretching and squeezing. Therefore, the tendex lines must, in some very deep way, represent the warping of space and time. And indeed they do, as the mathematics of the Riemann tensor taught us.

Black holes are not the only objects that produce stretching and squeezing forces. Stars and planets and moons also produce them. In 1687 Isaac Newton discovered them in his own theory of gravity and used them to explain ocean tides.

The Moon's gravity pulls more strongly on the near face of the Earth than on the far face, Newton reasoned. And the direction of pull on the Earth's sides is slightly inward, because it is toward the Moon's center, a slightly different direction on the Earth's two sides. This is the usual viewpoint about the Moon's gravity depicted in Figure 4.8.

Now, the Earth does *not feel the average* of these gravitational pulls, because it is falling freely along its orbit.[4] (This is like the *Endurance*'s crew not feeling Gargantua's gravitational pull when they are in the *Endurance*, in its parking orbit above the black hole. They only feel centrifugal forces due to the *Endurance*'s rotation.) What the Earth *does* feel is the red-arrowed lunar pulls in the left

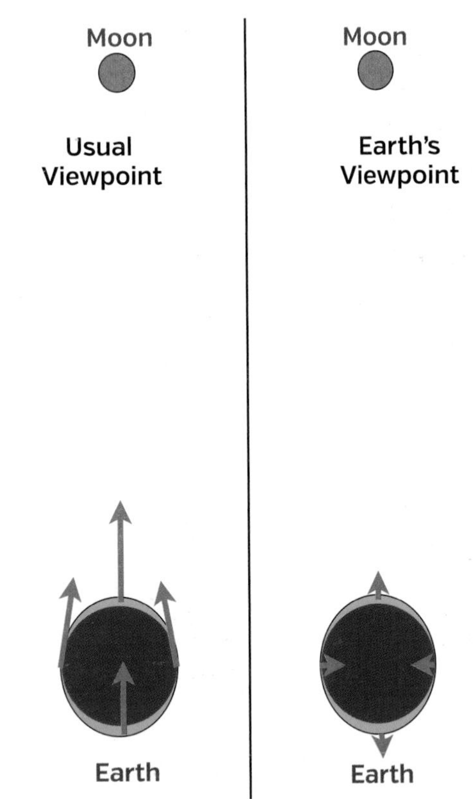

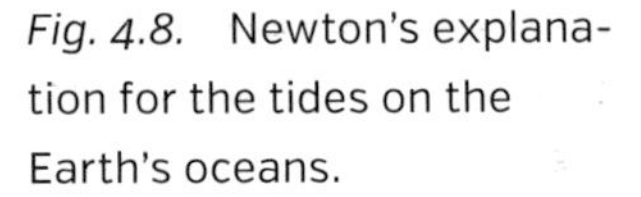
Fig. 4.8. Newton's explanation for the tides on the Earth's oceans.

4 In 1907, Einstein realized that if he were to fall, off the roof of his house for example, then as he fell he would feel no gravity. He called this the "happiest thought of my life," because it got him started on his quest to understand gravity, the quest that led to his concepts of warped time and space and the laws that govern the warping.

half of Figure 4.8, with their average subtracted away; that is, it feels a stretch toward and away from the Moon, and a squeeze on its lateral sides (right half of Figure 4.8). This is qualitatively the same as around a black hole (Figure 4.7).

These felt forces stretch the ocean away from the Earth's surface on the faces toward and away from the Moon, producing high tides there. And the felt forces squeeze the oceans toward the Earth's surface on the Earth's lateral sides, producing low tides there. As the Earth turns on its axis, one full turn each twenty-four hours, we see two high tides and two low tides. This was Newton's explanation of ocean tides, aside from a slight complication: The Sun's tidal gravity also contributes to the tides. Its stretch and squeeze get added to the Moon's stretch and squeeze.

Because of their role in ocean tides, these gravitational squeezing and stretching forces—the forces the Earth *feels*—are called tidal forces. To extremely high accuracy, these tidal forces, computed using Newton's laws of gravity, are the same as we compute using Einstein's relativistic laws. They must be the same, since the relativistic laws and the Newtonian laws always make the same predictions when gravity is weak and objects move at speeds much slower than light.

In the relativistic description of the Moon's tides (Figure 4.9), the tidal forces are produced by blue tendex lines that squeeze the Earth's lateral sides and red tendex lines that stretch toward and away from the Moon. This is just like a black hole's tendex lines (Figure 4.7). The Moon's tendex lines are visual embodiments of the Moon's warping of space and time. It is remarkable that a warping so tiny can produce forces big enough to cause the ocean tides!

On Miller's planet (Chapter 17) the tidal forces are enormously larger and are key to the huge waves that Cooper and his crew encounter.

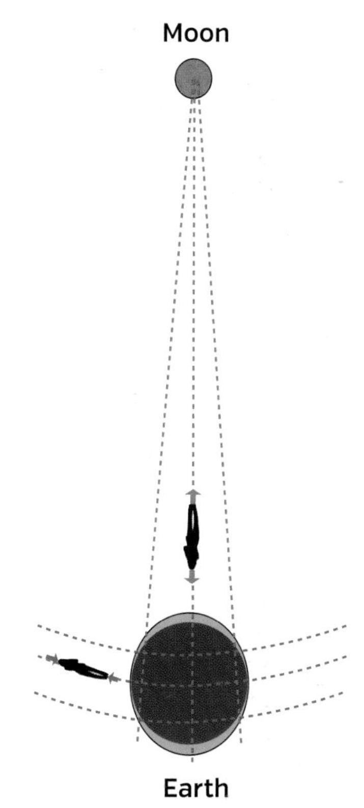

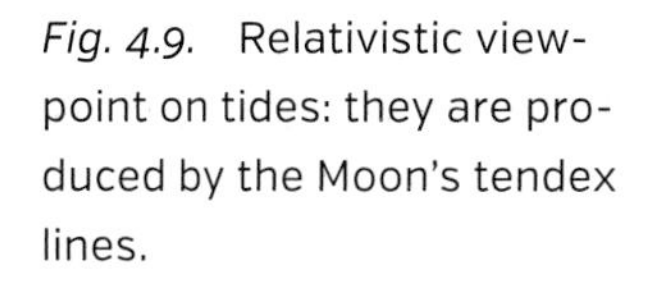

Fig. 4.9. Relativistic viewpoint on tides: they are produced by the Moon's tendex lines.

We now have three points of view on tidal forces:

- *Newton's viewpoint* (Figure 4.8): The Earth does not feel the Moon's full gravitational pull, but rather the full pull (which varies over the Earth) minus the average pull.
- *The tendex viewpoint* (Figure 4.9): The Moon's tendex lines stretch

and squeeze the Earth's oceans; also (Figure 4.7) a black hole's tendex lines stretch and squeeze the paths of planets and stars around the black hole.

- *The straightest-route viewpoint* (Figure 4.6): The paths of stars and planets around a black hole are the straightest routes possible in the hole's warped space and time.

Having three different viewpoints on the same phenomenon can be extremely valuable. Scientists and engineers spend most of their lives trying to solve puzzles. The puzzle may be how to design a spacecraft. Or it may be figuring out how black holes behave. Whatever the puzzle may be, if one viewpoint doesn't yield progress, another viewpoint may. Peering at the puzzle first from one viewpoint and then from another can often trigger new ideas. This is what Professor Brand does, in *Interstellar,* when trying to understand and harness gravitational anomalies (Chapters 24 and 25). This is what I've spent most of my adult life doing.

Black Holes

Ⓣ

The black hole Gargantua plays a major role in *Interstellar.* Let's look at the basic facts about black holes in this chapter and then focus on Gargantua in the next.

First, a weird claim: *Black holes* are made from warped space and warped time. Nothing else—no matter whatsoever.

Now some explanation.

Ant on a Trampoline: A Black Hole's Warped Space

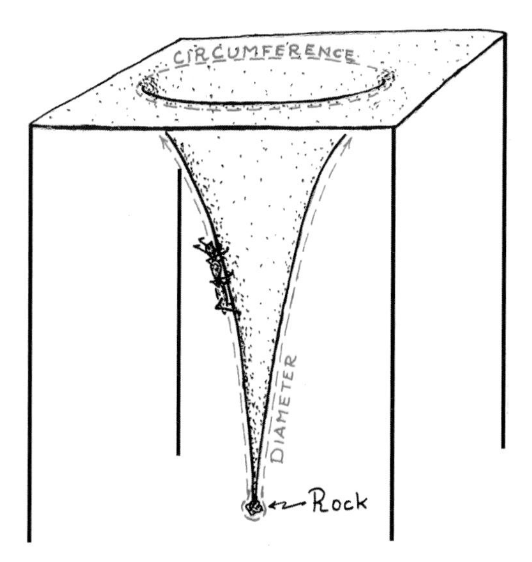

Fig. 5.1. An ant on a warped trampoline. *[My own hand sketch.]*

Imagine you're an ant and you live on a child's trampoline—a sheet of rubber stretched between tall poles. A heavy rock bends the rubber downward, as shown in Figure 5.1. You're a blind ant, so you can't see the poles or the rock or the bent rubber sheet. But you're a smart ant. The rubber sheet is your entire universe, and you suspect it's warped. To determine its shape, you walk around a circle in the upper region measuring its circumference, and then walk through the center from one side of the circle to the other, measuring its diameter. If your universe were flat, then the circumference would be π = 3.14159... times the diameter. But the circumference, you discover, is far smaller than the diameter. Your universe, you conclude, is highly warped!

Space around a nonspinning black hole has the same warping as the trampoline: Take an equatorial slice through the black hole. This is a two-dimensional surface. As seen from the bulk, this surface is warped in the same manner as the trampoline. Figure 5.2 is the same as Figure 5.1, with the ant and poles removed and the rock replaced by a *singularity* at the black hole's center.

The singularity is a tiny region where the surface forms a point and thus is "infinitely warped," and where, it turns out, tidal gravitational forces are infinitely strong, so matter as we know it gets stretched and squeezed out of existence. In chapters 26, 28, and 29, we see that Gargantua's singularity is somewhat different from this one, and why.

For the trampoline, the warping of space is produced by the rock's weight. Similarly, one might suspect, the black hole's space warp is produced by the singularity at its center. Not so. In fact, the hole's space is warped by the enormous energy of its warping. Yes, that's what I meant to say. If this seems a bit circular to you, well, it is, but it has deep meaning.

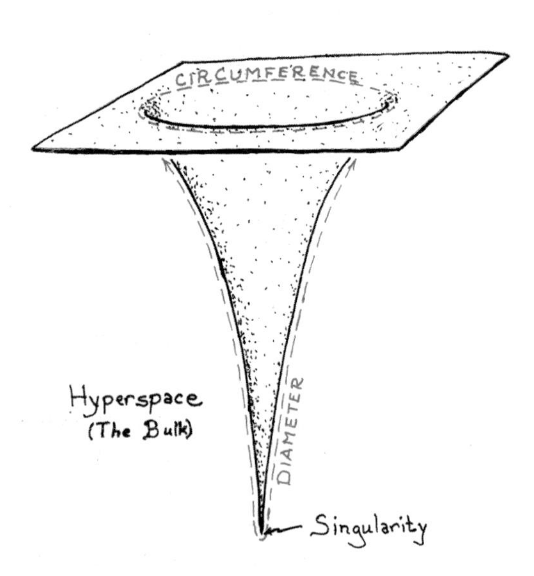

Fig. 5.2. The warped space inside and around a black hole, as seen from the bulk. *[My own hand sketch.]*

Just as it requires a lot of energy to bend a stiff bow in preparation for shooting an arrow, so it requires a lot of energy to bend space; to warp it. And just as the bending energy is stored in the bent bow (until the string is released and feeds the bow's energy into the arrow), so the warping energy is stored in the black hole's warped space. And for a black hole, that energy of warping is so great that it generates the warping.

Warping begets warping in a nonlinear, self-bootstrapping manner. This is a fundamental feature of Einstein's relativistic laws, and so different from everyday experience. It's somewhat like a hypothetical science-fiction character who goes backward in time and gives birth to herself.

This warping-begets-warping scenario does not happen in our solar system hardly at all. Throughout our solar system the space warps are so weak that their energy is minuscule, far too small to produce much bootstrapped warping. Almost all the space warping in our solar system is produced directly by matter—the Sun's matter, the Earth's matter, the matter of the other planets—by contrast with a black hole where the warping is fully responsible for the warping.

Event Horizon and Warped Time

When you first hear mention of a black hole, you probably think of its trapping power as depicted in Figure 5.3, not its warped space.

If I fall into a black hole carrying a microwave transmitter, then once I pass through the hole's *event horizon*, I'm pulled inexorably on downward, into the hole's singularity. And any signals I try to transmit in any manner whatsoever get pulled down with me. Nobody above the horizon can ever see the signals I send after I cross the horizon. My signals and I are trapped inside the black hole. (See Chapter 28 for how this plays out in *Interstellar*.)

This trapping is actually caused by the hole's time warp. If I hover above the black hole, supporting myself by the blast of a rocket engine, then the closer I am to the horizon, the more slowly my time flows. At the horizon itself, time slows to a halt and, therefore, according to Einstein's law of time warps, I must experience an infinitely strong gravitational pull.

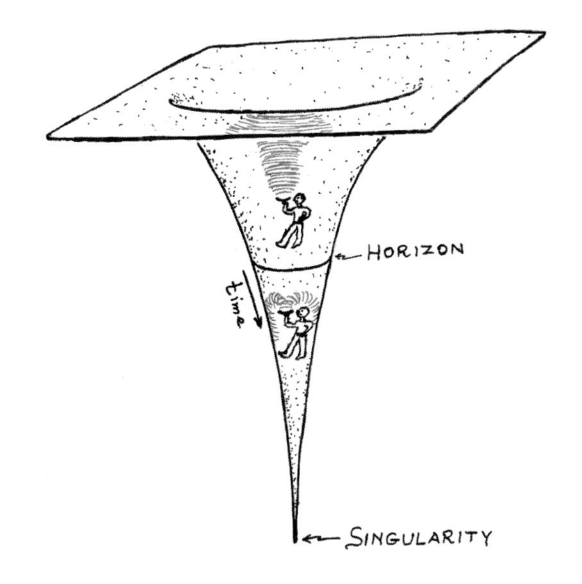

Fig. 5.3. Signals I send after crossing the event horizon can't get out. *Note:* Because one space dimension is removed from this diagram, I am a two-dimensional Kip, sliding down the warped two-dimensional surface, part of our brane. *[My own hand sketch.]*

What happens inside the event horizon? Time is so extremely warped there that it flows in a direction you would have thought was spatial: it flows downward toward the singularity. That downward

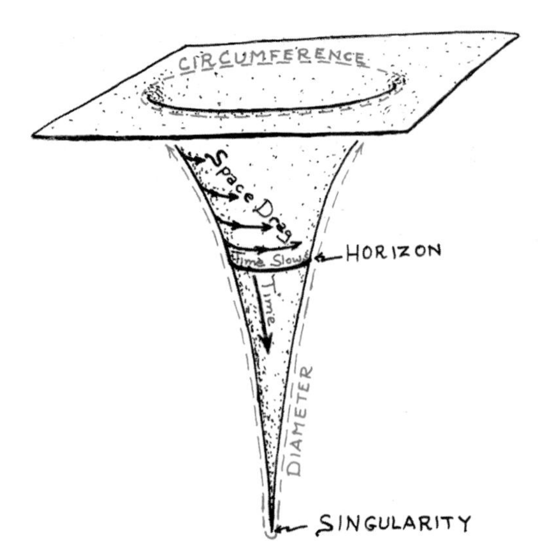

Fig. 5.4. Space around a spinning black hole is dragged into whirling motion. *[My own hand sketch.]*

flow, in fact, is why nothing can escape from a black hole. Everything is drawn inexorably toward the future,[1] and since the future inside the hole is downward, away from the horizon, nothing can escape back upward, through the horizon.

Space Whirl

Black holes can spin, just as the Earth spins. A spinning hole drags space around it into a vortex-type, whirling motion (Figure 5.4). Like the air in a tornado, space whirls fastest near the hole's center, and the whirl slows as one moves outward, away from the hole. Anything that falls toward the hole's horizon gets dragged, by the whirl of space, into a whirling motion around and around the hole, like a straw caught and dragged by a tornado's wind. Near the horizon there is no way whatsoever to protect oneself against this whirling drag.

Precise Depiction of the Warped Space and Time Around a Black Hole

These three aspects of spacetime warping—the warp of space, the slowing and distortion of time, and the whirl of space—are all described by mathematical formulas. These formulas have been deduced from Einstein's relativistic laws, and their precise predictions are depicted quantitatively in Figure 5.5 (by contrast with Figures 5.1–5.4, which were only qualitative).

The warped shape of the surface in Figure 5.5 is precisely what we would see from the bulk, when looking at the hole's equatorial plane. The colors depict the slowing of time as measured by someone who

1 If it is possible to go backward in time, you can only do so by traveling outward in space and then returning to your starting point before you left. You *cannot* go backward in time at some fixed location, while watching others go forward in time there. More on this in Chapter 30.

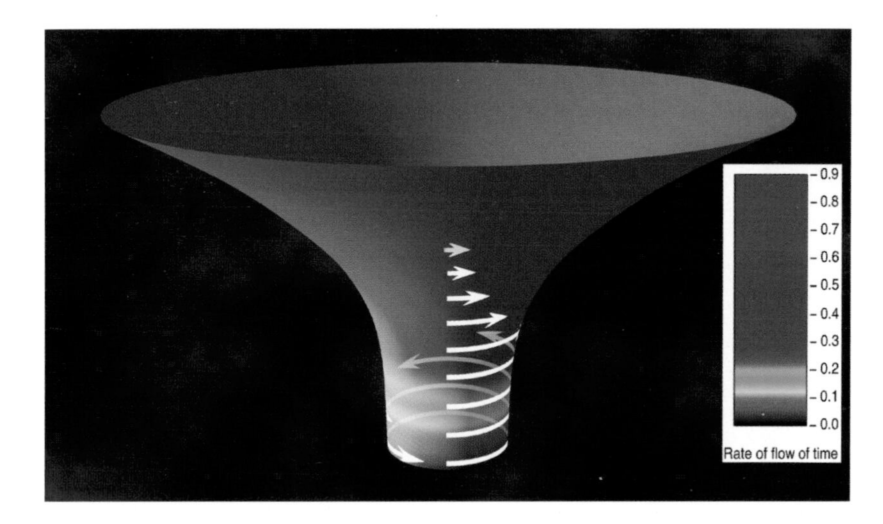

Fig. 5.5. Precise depiction of the warped space and time around a rapidly spinning black hole: one that spins at 99.8 percent of the maximum possible rate. *[Drawing by Don Davis based on a sketch by me.]*

hovers at a fixed height above the horizon. At the transition from blue to green, time flows 20 percent as fast as it flows far from the hole. At the transition from yellow to red, time is slowed to 10 percent of its normal rate far away. And at the black circle, the bottom of the surface, time slows to a halt. This is the event horizon. It is a circle, not a sphere, because we are looking only at the equatorial plane, only at two dimensions of our universe (of our brane). If we were to restore the third space dimension, the horizon would become a flattened sphere: a spheroid. The white arrows depict the rate at which space whirls around the black hole. The whirl is fast at the horizon, and decreases as we climb upward in a spacecraft.

In the fully accurate Figure 5.5, I don't depict the hole's interior. We'll get to that later, in Chapters 26 and 28.

The warping in Figure 5.5 is the essence of a black hole. From its details, expressed mathematically, physicists can deduce everything about the black hole, except the nature of the singularity at its center. For the singularity, they need the ill-understood laws of quantum gravity (Chapters 26).

A Black Hole's Appearance from Inside Our Universe

We humans are confined to our brane. We can't escape from it, into the bulk (unless an ultra-advanced civilization gives us a ride in a tes-

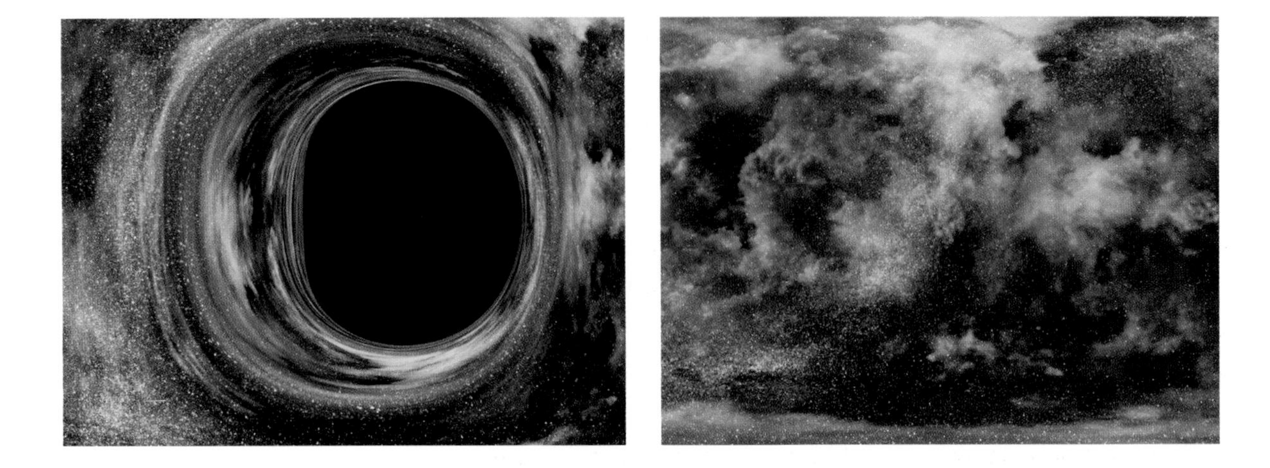

Fig. 5.6. A fast-spinning black hole (*left*) moving in front of the star field shown on the right. *[From a simulation for this book by the Double Negative visual-effects team.]*

seract or some such vehicle, as they do for Cooper in *Interstellar*; see Chapter 29). Therefore, we can't see a black hole's warped space, as depicted in Figure 5.5. The black-hole funnels and whirlpools so often shown in movies, for example, Disney Studios' 1979 movie *The Black Hole*, would never be seen by any creature that lives in our universe.

Interstellar is the first Hollywood movie to depict a black hole correctly, in the manner that humans would actually see and experience it. Figure 5.6 is a example, not taken from the movie. The black hole casts a black shadow on the field of stars behind it. Light rays from the stars are bent by the hole's warped space; they are *gravitationally lensed*, producing a concentric pattern of distortion. Light rays coming to us from the shadow's left edge move in the same direction as the hole's whirling space. The space whirl gives them a boost, letting them escape from closer to the horizon than light rays on the shadow's right edge, which struggle against the whirl of space. That's why the shadow is flattened on the left and bulges out on the right. In Chapter 8 I talk more about this and other aspects of what a black hole really looks like, as seen up close in our universe, in our *brane*.

How Do We Know This Is *True*?

Einstein's relativistic laws have been tested to high precision. I'm convinced they are right, except when they confront quantum physics. For a big black hole like *Interstellar*'s Gargantua, quantum physics is relevant only near its center, in its singularity. So if black holes exist at all in our universe, they must have the properties that Einstein's relativistic laws dictate, the properties I described above.

These properties and others have been deduced from Einstein's equations by a large number of physicists standing intellectually on each others' shoulders (Figure 5.7); most importantly, Karl Schwarzschild, Roy Kerr, and Stephen Hawking. In 1915, shortly before his tragic death on World War I's German/Russian front, Schwarzschild deduced the details of the warped spacetime around a nonspinning black hole. In physicists' jargon, those details are called the "Schwarzschild metric." In 1963, Kerr (a New Zealand mathematician) did the same for a spinning black hole: he deduced the spinning hole's "Kerr metric." And in the early 1970s Stephen Hawking and others deduced a set of laws that black holes must obey when they swallow stars, collide and merge, and feel the tidal forces of other objects.

Black holes surely do exist. Einstein's relativistic laws insist that, when a massive star exhausts the nuclear fuel that keeps it hot, then the star must implode. In 1939, J. Robert Oppenheimer and his student Hartland Snyder used Einstein's laws to discover that, if the implosion is precisely spherical, the imploding star *must* create a black hole around itself, and then create a singularity at the hole's center, and then get swallowed into the singularity. No matter is left behind. None whatsoever. The resulting black hole is made entirely from warped space and time. Over the decades since 1939, physicists using Einstein's laws have shown that if the imploding star is deformed and spinning, it will also produce a black hole. Computer simulations reveal the full details.

Astronomers have seen compelling evidence for many black holes in our universe. The most beautiful example is a massive black hole at

Fig 5.7. Black-hole scientists. *Left to right*: Karl Schwarzschild (1873–1916), Roy Kerr (1934–), Stephen W. Hawking (1942–), J. Robert Oppenheimer (1904–1967), and Andrea Ghez (1965–).

the center of our Milky Way galaxy. Andrea Ghez of UCLA, with a small group of astronomers that she leads, has monitored the motions of stars around that black hole (Figure 5.8). Along each orbit, the dots are the star's position at times separated by one year. I marked the black hole's location by a white, five-pointed symbol. From the stars' observed motions, Ghez has deduced the strength of the hole's gravity. Its gravitational pull, at a fixed distance, is 4.1 million times greater than the Sun's pull at that distance. This means the black hole's mass is 4.1 million times greater than the Sun's!

Figure 5.9 shows where this black hole is on the night sky in summer. It is to the lower right of the constellation Sagittarius, the teapot, at the × labeled "Galactic Center."

A massive black hole inhabits the core of nearly every big galaxy in our universe. Many of these are as heavy as Gargantua (100 million Suns), or even heavier. The heaviest yet measured is 17 billion times more massive than the Sun; it resides at the center of a galaxy whose

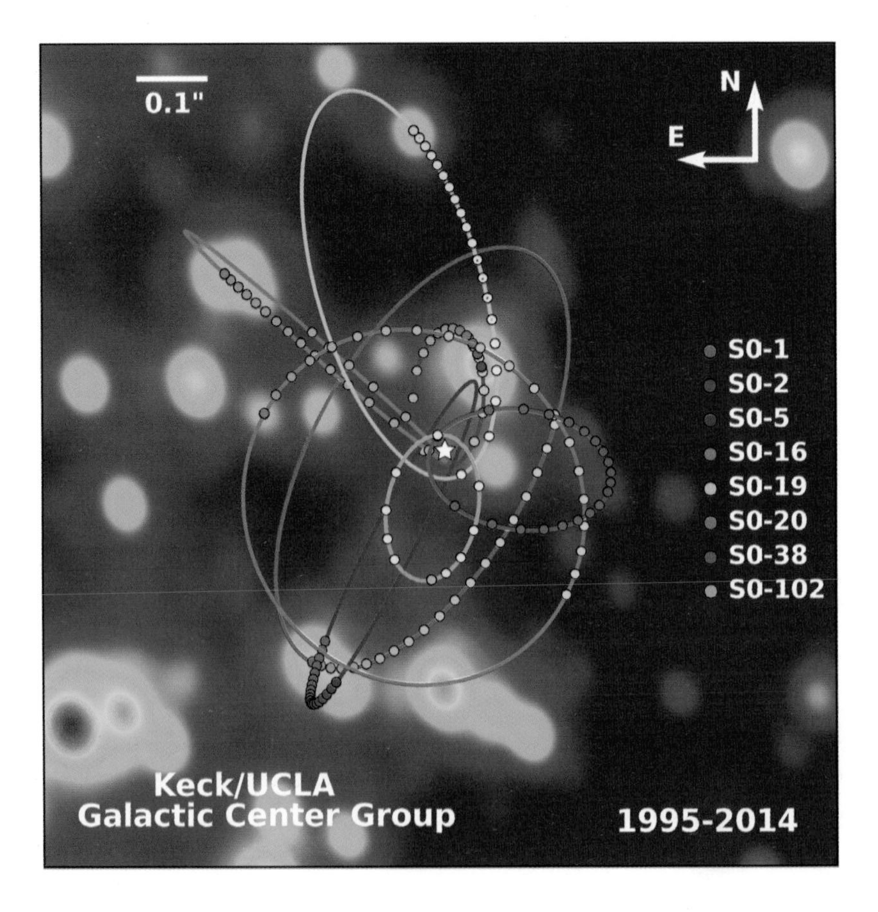

Fig. 5.8. Observed orbits of stars around the massive black hole at the center of our Milky Way galaxy, as measured by Andrea Ghez and colleagues.

Fig. 5.9. The location of our galaxy's center on the sky. A giant black hole resides there.

name is NGC1277, 250 million light-years from Earth—roughly a tenth of the way to the edge of the visible universe.

Inside our own galaxy, there are roughly 100 million smaller black holes: holes that typically are between about three and thirty times as heavy as the Sun. We know this not because we've seen evidence for all these, but because astronomers have made a census of heavy stars that will become black holes when they exhaust their nuclear fuel. From that census, astronomers have inferred how many such stars have already exhausted their fuel and become black holes.

So black holes are ubiquitous in our universe. Fortunately, there are none in our solar system. If there were, the hole's gravity would wreak havoc with the Earth's orbit. The Earth would be thrown close to the Sun where it boils, or far from the Sun where it freezes, or even out of the solar system or into the black hole. We humans would survive for no more than a year or so!

Astronomers estimate that the nearest black hole to Earth is roughly 300 light-years away: a hundred times farther than the nearest star (other than the Sun), Proxima Centauri.

Now armed with a basic understanding of the universe, fields, warped time and space and especially black holes, we are ready, at last, to explore *Interstellar*'s Gargantua.

II

GARGANTUA

6

Gargantua's Anatomy

If we know the mass of a black hole and how fast it spins, then from Einstein's relativistic laws we can deduce all the hole's other properties: its size, the strength of its gravitational pull, how much its event horizon is stretched outward near the equator by centrifugal forces, the details of the gravitational lensing of objects behind it. Everything.

This is amazing. So different from everyday experience. It is as though knowing my weight and how fast I can run, you could deduce everything about me: the color of my eyes, the length of my nose, my IQ, . . .

John Wheeler (my mentor, who gave "black holes" their name) has described this by the phrase "A black hole has no hair"—no extra, *independent* properties beyond its mass and its spin. Actually, he should have said, "A black hole has only two hairs, from which you can deduce everything else about it," but that's not as catchy as "no hair," which quickly became embedded in black-hole lore and scientists' lexicon.[1]

1 The literal French translation of "a black hole has no hair" is so obscene that French publishers resisted it vigorously, to no avail.

From the properties of Miller's planet, as depicted in *Interstellar,* a physicist who knows Einstein's relativistic laws can deduce Gargantua's mass and spin, and thence all else about it. Let's see how this works.[2]

Gargantua's Mass

Ⓣ

Miller's planet (which I talk about at length in Chapter 17) is about as close to Gargantua as it can possibly be and still survive. We know this because the crew's extreme loss of time can only occur very near Gargantua.

At so close a distance, Gargantua's tidal gravitational forces (Chapter 4) are especially strong. They stretch Miller's planet toward and away from Gargantua and squeeze the planet's sides (Figure 6.1).

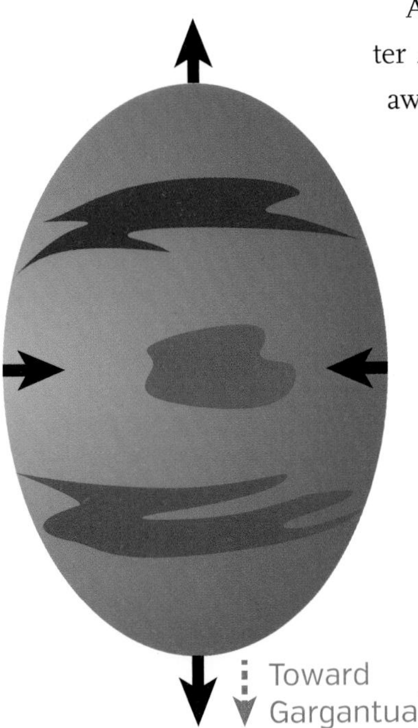

Fig. 6.1. Gargantua's tidal gravitational forces stretch and squeeze Miller's planet.

The strength of this stretch and squeeze is inversely proportional to the square of Gargantua's mass. Why? The greater Gargantua's mass, the greater its circumference, and therefore the more similar Gargantua's gravitational forces are on the various parts of the planet, which results in weaker tidal forces. (See Newton's viewpoint on tidal forces; Figure 4.8.) Working through the details, I conclude that Gargantua's mass must be at least 100 million times bigger than the Sun's mass. If Gargantua were less massive than that, it would tear Miller's planet apart!

In all my science interpretations of what happens in *Interstellar,* I assume that this actually *is* Gargantua's mass: 100 million Suns.[3] For example, I assume this mass in Chapter 17, when explaining how Gargantua's tidal forces could produce the giant water waves that inundate the Ranger on Miller's planet.

The circumference of a black hole's event horizon is proportional to the hole's mass. For Gargantua's 100 million solar masses, the hori-

2 For some quantitative details, see *Some Technical Notes,* at the end of this book.

3 A more reasonable value might be 200 million times the Sun's mass, but I want to keep the numbers simple and there's a lot of slop in this one, so I chose 100 million.

zon circumference works out to be approximately the same as the Earth's orbit around the Sun: about *1 billion kilometers*. That's big! After consulting with me, that's the circumference assumed by Paul Franklin's visual-effects team, when producing the images in *Interstellar.*

Physicists attribute to a black hole a radius equal to its horizon's circumference divided by 2π (about 6.28). Because of the extreme warping of space inside the black hole, this is not the hole's true radius. Not the true distance from the horizon to the hole's center, as measured in our universe. But it *is* the event horizon's radius (half its diameter) as measured in the bulk; see Figure 6.3 below. Gargantua's radius, in this sense, is about 150 million kilometers, the same as the radius of the Earth's orbit around the Sun.

Gargantua's Spin

When Christopher Nolan told me how much slowing of time he wanted on Miller's planet, *one hour there is seven years back on Earth*, I was shocked. I didn't think that possible and I told Chris so. "It's non-negotiable," Chris insisted. So, not for the first time and also not the last, I went home, thought about it, did some calculations with Einstein's relativistic equations, and found a way.

I discovered that, if Miller's planet is about as near Gargantua as it can get without falling in,[4] and if Gargantua is spinning fast enough, then Chris's one-hour-in-seven-years time slowing is possible. But Gargantua has to spin *awfully fast*.

There is a maximum spin rate that any black hole can have. If it spins faster than that maximum, its horizon disappears, leaving the singularity inside it wide open for all the universe to see; that is, making it *naked*—which is probably forbidden by the laws of physics (Chapter 26).

I found that Chris's huge slowing of time requires Gargantua to spin almost as fast as the maximum: less than the maximum by about

4 See Figure 17.2 and the discussion of it in Chapter 17.

one part in 100 trillion.[5] In most of my science interpretations of *Interstellar,* I assume this spin.

The crew of the *Endurance* could measure the spin rate directly by watching from far, far away as the robot TARS falls into Gargantua (Figure 6.2).[6] As seen from afar, TARS never crosses the horizon (because signals he sends after crossing can't get out of the black hole). Instead, TARS' infall appears to slow down, and he appears to hover just above the horizon. And as he hovers, Gargantua's whirling space sweeps him around and around Gargantua, as seen from afar. With Garantua's spin very near the maximum possible, TARS' orbital period is about one hour, as seen from afar.

You can do the math yourself: the orbital distance around Gargantua is a billion kilometers and TARS covers that distance in one hour,

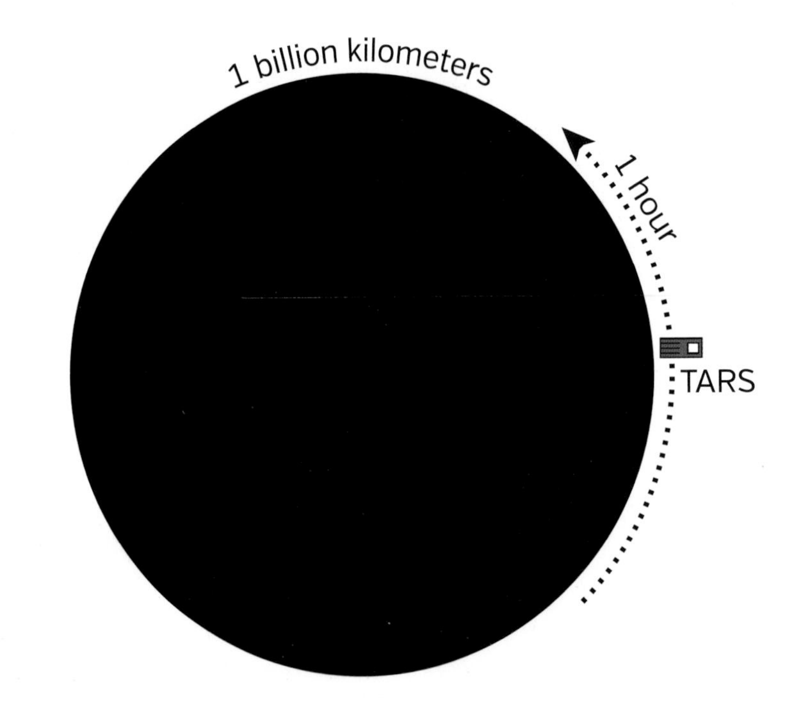

Fig. 6.2. TARS, falling into Gargantua, is dragged around the hole's billion-kilometer circumference once each hour, as seen from afar.

5 In other words, its spin is the maximum minus 0.00000000000001 of the maximum.

6 When TARS falls in, the *Endurance* is not far, far away, but rather is on the critical orbit, quite near the horizon, whirling around the hole nearly as fast as TARS; so Amelia Brand, in the *Endurance,* does not see TARS swept around at high speed. For more on this, see Chapter 27.

so his speed as measured from afar is about a billion kilometers per hour, which is approximately the speed of light! If Gargantua were spinning faster than the maximum, TARS would whip around faster than the speed of light, which violates Einstein's speed limit. This is a heuristic way to understand why there is a maximum possible spin for any black hole.

In 1975, I discovered a mechanism by which Nature protects black holes from spinning faster than the maximum: When it gets close to the maximum spin, a black hole has difficulty capturing objects that orbit in the same direction as the hole rotates and that therefore, when captured, increase the hole's spin. But the hole easily captures things that orbit opposite to its spin and that, when captured, slow the hole's spin. Therefore, the spin is easily slowed, when it gets close to the maximum.

In my discovery, I focused on a disk of gas, somewhat like Saturn's rings, that orbits in the same direction as the hole's spin: an *accretion disk* (Chapter 9). Friction in the disk makes the gas gradually spiral into the black hole, spinning it up. Friction also heats the gas, making it radiate photons. The whirl of space around the hole grabs those photons that travel in the same direction as the hole spins and flings them away, so they can't get into the hole. By contrast, the whirl grabs photons that are trying to travel opposite to the spin and sucks them into the hole, where they slow the spin. Ultimately, when the hole's spin reaches 0.998 of the maximum, an equilibrium is reached, with spin-down by the captured photons precisely counteracting spin-up by the accreting gas. This equilibrium appears to be somewhat robust. In most astrophysical environments I expect black holes to spin no faster than about 0.998 of the maximum.

However, I can imagine situations—very rare or never in the real universe, but possible nevertheless—where the spin gets much closer to the maximum, even as close as Chris requires to produce the slowing of time on Miller's planet, a spin one part in 100 trillion less than the maximum spin. Unlikely, but possible.

This is common in movies. To make a great film, a superb filmmaker often pushes things to the extreme. In science fantasy films such as *Harry Potter*, that extreme is far beyond the bounds of the scientifically possible. In science fiction, it's generally kept in the realm of the possible. That's the main distinction between science fantasy and science fiction. *Interstellar* is science fiction, not fantasy. Gargantua's ultrafast spin is scientifically possible.

Gargantua's Anatomy

Ⓣ

Having determined Gargantua's mass and spin, I used Einstein's equations to compute its anatomy. As in the previous chapter, here we focus solely on the external anatomy, leaving the interior (especially Gargantua's singularities) for Chapters 26 and 28.

In the top picture in Figure 6.3, you see the shape of Gargantua's equatorial plane as viewed from the bulk. This is like Figure 5.5, but because Gargantua's spin is much closer to the maximum possible (one part in 100 trillion contrasted with two parts in a thousand in

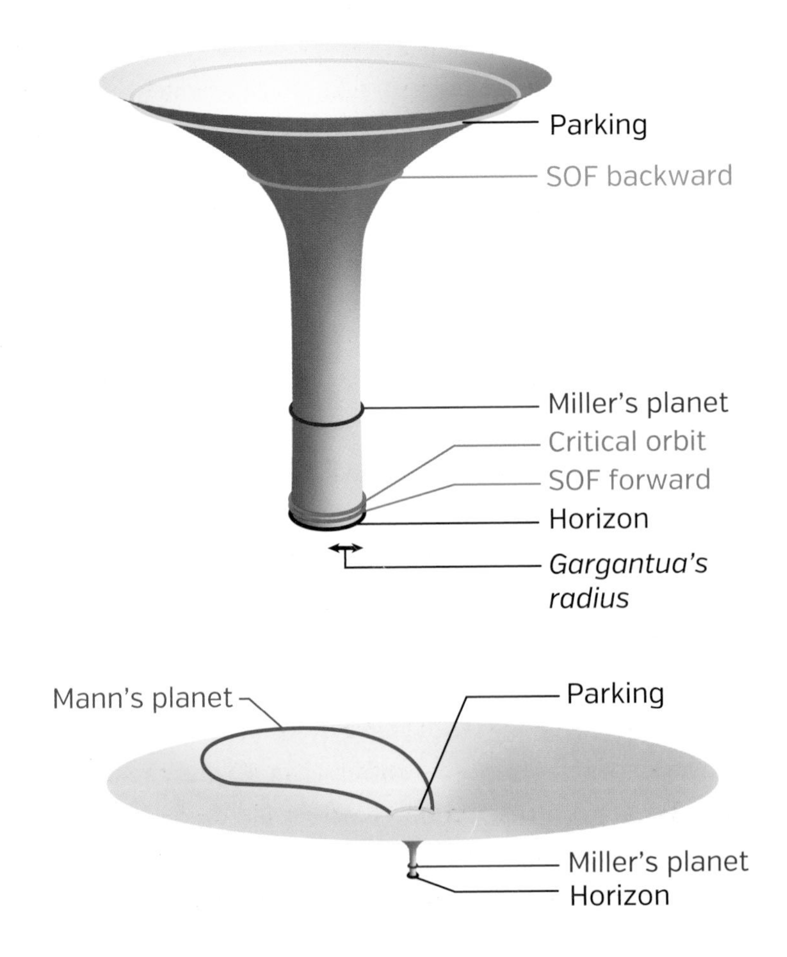

Fig. 6.3. Gargantua's anatomy, when its spin is only one part in 100 trillion smaller than the maximum possible, as is required to get the extreme slowing of time on Miller's planet.

Figure 5.5), Gargantua's throat is far longer. It extends much farther downward before reaching the horizon. The region near the horizon, as seen from the bulk, looks like a long cylinder. The length of the cylindrical region is about two horizon circumferences, that is, 2 billion kilometers.

The cylinder's cross sections are circles in the diagram, but if we were to restore the third dimension of our brane by moving out of Gargantua's equatorial plane, the cross sections would become flattened spheres (spheroids).

On Gargantua's equatorial plane I marked several special locations that occur in my science interpretations of *Interstellar*: Gargantua's event horizon (black circle), the critical orbit from which Cooper and TARS fall into Gargantua near the end of the movie (green circle; Chapter 27), the orbit of Miller's planet (blue circle; Chapter 17), the orbit in which the *Endurance* is parked while the crew visit Miller's planet (yellow circle), and a segment of the nonequatorial orbit of Mann's planet, projected into the equatorial plane (purple circle). The outer part of Mann's orbit is so far away from Gargantua (600 Gargantua radii or more; Chapter 19) that I had to redraw the picture on a much larger scale to fit it in (bottom picture), and, even so, I didn't do it honestly: I only put the outer part at 100 Gargantua radii instead of 600 as I should. The red circles are labeled "SOF" for "shell of fire"; see below.

How did I come up with these locations? I use the parking orbit as an illustration here and discuss the others later. In the movie, Cooper describes the parking orbit this way: "So we track a wider orbit of Gargantua, parallel with Miller's planet but a little further out." And he wants it to be far enough from Gargantua to be "out of the time shift," that is, far enough from Gargantua that the slowing of time compared to Earth is very modest. This motivated my choice of five Gargantua radii (yellow circle in Figure 6.3). The time for the Ranger to travel from this parking orbit to Miller's planet, two and a half hours, reinforced my choice.

But there was a problem with this choice. At this distance, Gargantua would look huge; it would subtend about 50 degrees on the *Endurance*'s sky. Truly awe inspiring, but undesirable for so early in the movie! So Chris and Paul chose to make Gargantua look much smaller at the parking orbit: about two and a half degrees, which is five times the size of the Moon as seen from Earth—still impressive but not overwhelmingly so.

The Shell of Fire

Ⓣ

Gravity is so strong near Gargantua, and space and time are so warped, that light (photons) can be trapped in orbits outside the horizon, traveling around and around the hole many times before escaping. These trapped orbits are *unstable* in the sense that the photons always escape from them, eventually. (By contrast, photons caught inside the horizon can never get out.)

I like to call this temporarily trapped light the "shell of fire." This fire shell plays an important role in the computer simulations (Chapter 8) that underlie Gargantua's visual appearance in *Interstellar.*

For a *non*spinning black hole, the shell of fire is a sphere, one with circumference 1.5 times larger than the horizon's circumference. The trapped light travels around and around this sphere on great circles (like the lines of constant longitude on the Earth); and some of it leaks into the black hole, while the rest leaks outward, away from the hole.

When a black hole is spun up, its shell of fire expands outward and inward, so it occupies a finite volume rather than just the surface of a sphere. For Gargantua, with its huge spin, the shell of fire in the equatorial plane extends from the bottom red circle of Figure 6.3 to the upper red circle. The shell of fire has expanded to encompass Miller's planet and the critical orbit, and much, much more! The bottom red circle is a light ray (a photon orbit) that moves around and around Gargantua in the same direction as Gargantua spins (the *forward* direction). The upper red circle is a photon orbit that moves in the opposite direction to Gargantua's spin (the *backward* direction). Evidently, the whirl of space enables the forward light to be much closer to the horizon without falling in than the backward light. What a huge effect the space whirl has!

The region of space occupied by the shell of fire above and below the equatorial plane is depicted in Figure 6.4. It is a large, annular

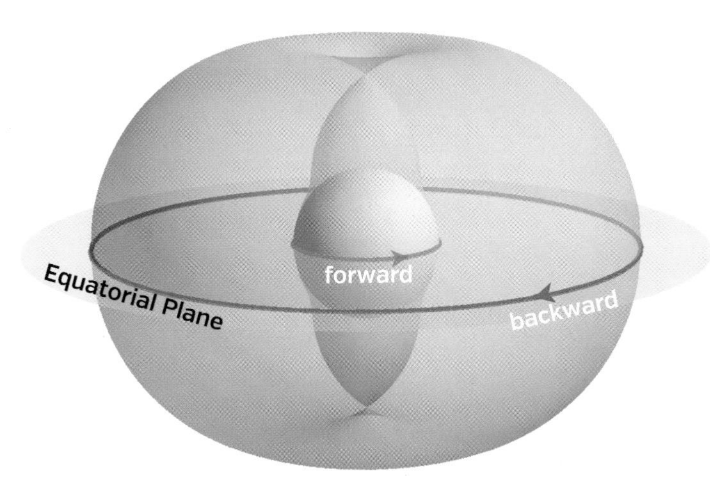

Fig. 6.4. The annular region around Gargantua, occupied by the shell of fire.

region. I omit the warping of space from this picture; it would get in the way of showing the shell of fire's full three dimensions.

Figure 6.5 shows some examples of photon orbits (light rays) trapped, temporarily, in the shell of fire.

The black hole is at the center of each of these orbits. The leftmost orbit winds around and around the equatorial region of a small sphere, traveling always forward, in the same direction as Gargantua's spin. It is nearly the same as the bottom (inner) red orbit in Figures 6.3 and 6.4. The next orbit in Figure 6.5 winds around a slightly larger sphere, in a nearly polar and slightly forward direction. The third orbit is larger still, but backward and nearly polar. The fourth is very nearly equatorial and backward, that is, nearly the upper (outer) red equatorial orbit of Figures 6.3 and 6.4. These orbits are actually inside each other; I pulled them apart so they are easier to see.

Some photons that are temporarily trapped in the shell of fire escape outward; they spiral away from Gargantua. The rest escape

Fig. 6.5. Examples of light rays (photon orbits) temporarily trapped in the shell of fire, as computed using Einstein's relativistic equations.

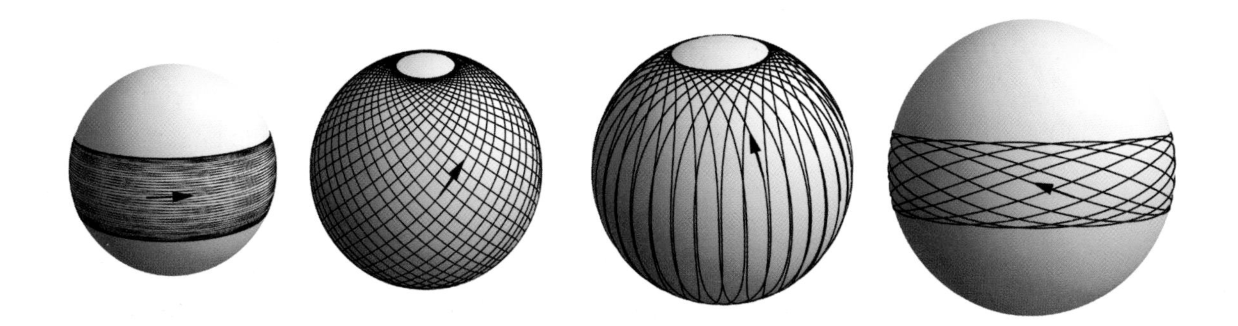

spiraling inward; they spiral toward Gargantua and plunge through its horizon. The nearly trapped but escaping photons have a big impact on Gargantua's visual appearance in *Interstellar.* They mark the edge of Gargantua's shadow as seen by the *Endurance*'s crew, and they produce a thin bright line along the shadow's edge: a "ring of fire" (Chapter 8).

7

Gravitational Slingshots

Ⓣ

Navigating a spacecraft near Gargantua is hard because the speeds are very high. To survive, a planet or star or spacecraft must counteract Gargantua's huge gravity with a comparably huge centrifugal force. This means it must move at very high speed. Near the speed of light, it turns out. In my science interpretation of *Interstellar*, the *Endurance*, parked at five Gargantua radii while the crew visit Miller's planet, moves at one-third the speed of light: $c/3$, where c represents the speed of light. Miller's planet moves at 55 percent the speed of light, $0.55c$.

To reach Miller's planet from the parking orbit in my interpretation (Figure 7.1), the Ranger must slow its forward motion from $c/3$ to far less than that, so Gargantua's gravity can pull it downward. And when it reaches the vicinity of the planet, the Ranger must turn from downward to forward. And, having picked up far too much speed while falling, it must slow by about $c/4$ to reach the planet's $0.55c$ speed and rendezvous with it.

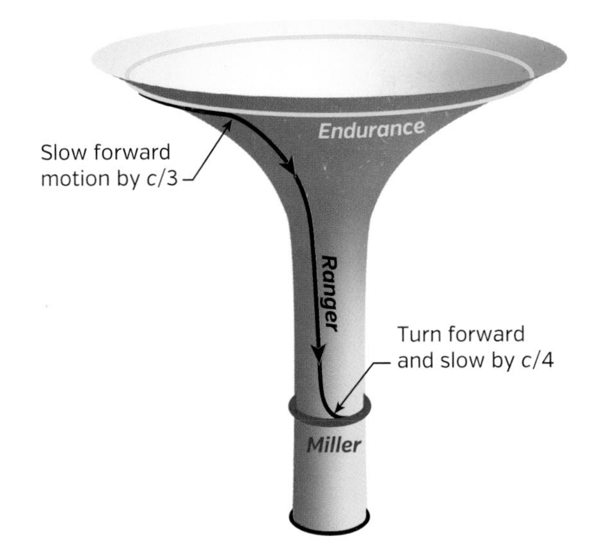

Fig. 7.1. The Ranger's trip to Miller's planet, in my interpretation of *Interstellar*.

What mechanism can Cooper, the Ranger's pilot, possibly use to produce these huge velocity changes?

Twenty-First-Century Technology

The required changes of velocity, roughly $c/3$, are 100,000 kilometers per second (per *second,* not per hour!).

By contrast, the most powerful rockets we humans have today can reach 15 kilometers per second: seven thousand times too slow. In *Interstellar,* the *Endurance* travels from Earth to Saturn in two years at an average speed of 20 kilometers per second, five thousand times too slow. The fastest that human spacecraft are likely to achieve in the twenty-first century, I think, is 300 kilometers per second. That would require a major R&D effort on nuclear rockets, but it is still three hundred times too slow for *Interstellar*'s needs.

Fortunately, Nature provides a way to achieve the huge speed changes, $c/3$, required in *Interstellar:* gravitational slingshots around black holes far smaller than Gargantua.

Slingshot Navigation to Miller's Planet

Stars and small black holes congregate around gigantic black holes like Gargantua (more on this in the next section). In my science interpretation of the movie, I imagine that Cooper and his team make a survey of all the small black holes orbiting Gargantua. They identify one that is well positioned to gravitationally deflect the Ranger from its near circular orbit and send it plunging downward toward Miller's planet (Figure 7.2). This gravity-assisted maneuver is called a "gravitational slingshot," and has often been used by NASA in the solar system—though with the gravity coming from planets rather than a black hole (see the end of the chapter).

This slingshot maneuver is not seen or discussed in *Interstellar,* but the next one *is* mentioned, by Cooper: "Look, I can swing around that *neutron star* to decelerate," he says. Deceleration is necessary because, having fallen under Gargantua's huge gravitational pull, from the *Endurance*'s orbit to Miller's orbit, the Ranger has acquired too much

Fig. 7.2. The Ranger performs a slingshot maneuver around a small black hole, deflecting it downward, toward Miller's planet.

speed; it is moving *c*/4 faster than Miller's planet. In Figure 7.3, the neutron star, traveling leftward relative to Miller's planet, deflects and slows the Ranger's motion so it can rendezvous gently with the planet.

Now, there is a feature of these slingshots that could be very unpleasant. Indeed, deadly: tidal forces (Chapter 4).

To change velocities by as much as *c*/3 or *c*/4, the Ranger must come close enough to the small black hole and neutron star to feel their intense gravity. At those close distances, if the deflector is a neutron star or is a black hole with radius less than 10,000 kilometers, the humans and the Ranger will be torn apart by tidal forces (Chapter 4). For the Ranger and humans to survive, the deflector must be a black hole at least 10,000 kilometers in size (about the size of the Earth).

Now, black holes that size *do* occur in Nature. They are called intermediate-mass black holes, or IMBHs, and despite their big size, they are tiny compared to Gargantua: ten thousand times smaller.

So Christopher Nolan should have used an Earth-sized IMBH to slow down the Ranger, not a neutron star. I discussed this with Chris early in his rewrites of Jonah's screenplay. After our discussion, Chris

Fig. 7.3. Slingshot around a neutron star enables the lander to rendezvous with Miller's planet.

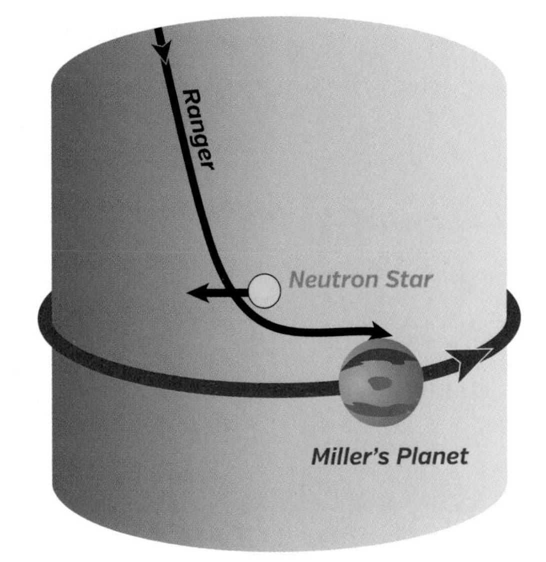

chose the neutron star. Why? Because he didn't want to confuse his mass audience by having more than one black hole in the movie. One black hole, one wormhole, and also a neutron star, along with *Interstellar*'s other rich science, all to be absorbed in a fast-paced two-hour film; that was all Chris thought he could get away with. Recognizing that strong gravitational slingshots *are* needed to navigate near Gargantua, Chris included one slingshot in Cooper's dialog, at the price of using a scientifically implausible deflector: the neutron star instead of a black hole.

Intermediate-Mass Black Holes in Galactic Nuclei

A 10,000-kilometer IMBH weighs about 10,000 solar masses. That's ten thousand times less than Gargantua, but a thousand times heavier than typical black holes. These are the deflectors Cooper needs.

Some IMBHs are thought to form in the cores of dense clusters of stars called globular clusters, and some of them are likely to find their way into the nuclei of galaxies, where gigantic black holes reside.

An example is Andromeda, the nearest large galaxy to our own (Figure 7.4), in whose nucleus lurks a Gargantua-sized black hole: 100 million solar masses. Huge numbers of stars are drawn into the

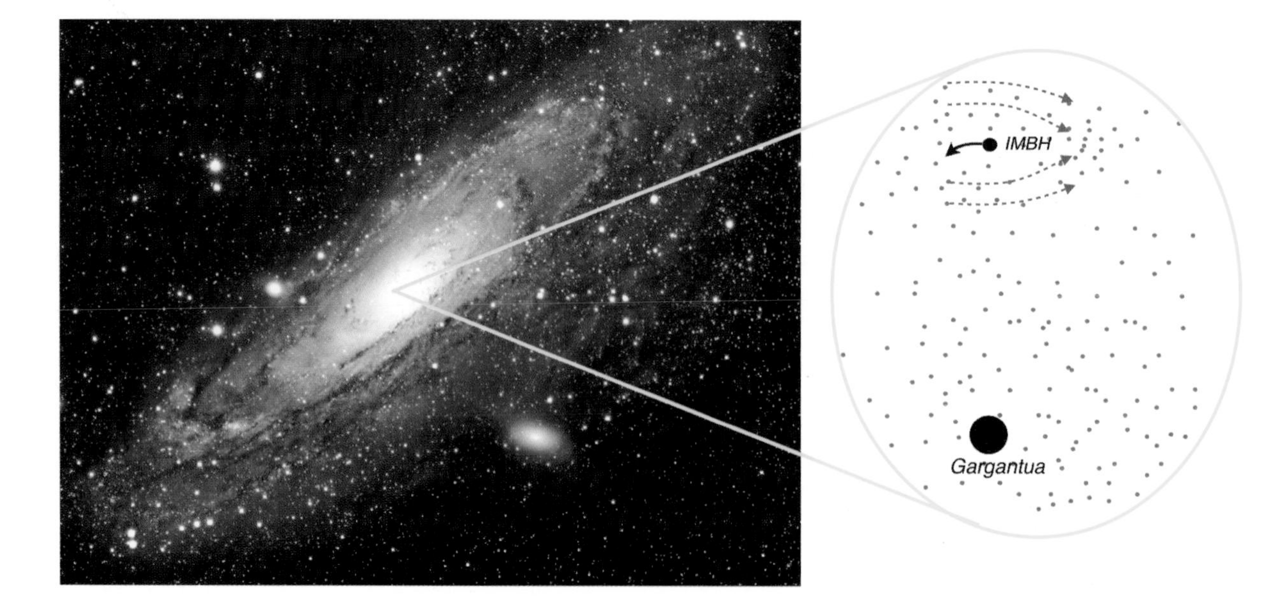

Fig. 7.4. *Left*: The Andromeda galaxy, which harbors a Gargantua-sized black hole. *Right*: The dynamical friction by which an IMBH will gradually slow down and sink into the vicinity of the gigantic black hole.

vicinity of such gigantic black holes; as many as a thousand stars per cubic light-year. When an IMBH passes through such a dense region, it gravitationally deflects the stars, creating a wake with enhanced density behind itself (Figure 7.4). The wake pulls on the IMBH gravitationally, slowing the IMBH down, a process called "dynamical friction." As the IMBH very gradually slows, it sinks deeper into the vicinity of the gigantic black hole. In this manner, Nature could provide Cooper, in my interpretation of *Interstellar*, with the IMBHs that he needs for his slingshots.[1]

Orbital Navigation by Ultra-Advanced Civilizations: A Digression

The orbits of planets and comets in our solar system are all ellipses to very high accuracy (Figure 7.5). Newton's laws of gravity guarantee and enforce this.

By contrast, around a gigantic, spinning black hole such as Gargantua, where Einstein's relativistic laws hold sway, the orbits are far more

Fig. 7.5. The orbits of planets, Pluto, and Halley's comet in our solar system are all ellipses.

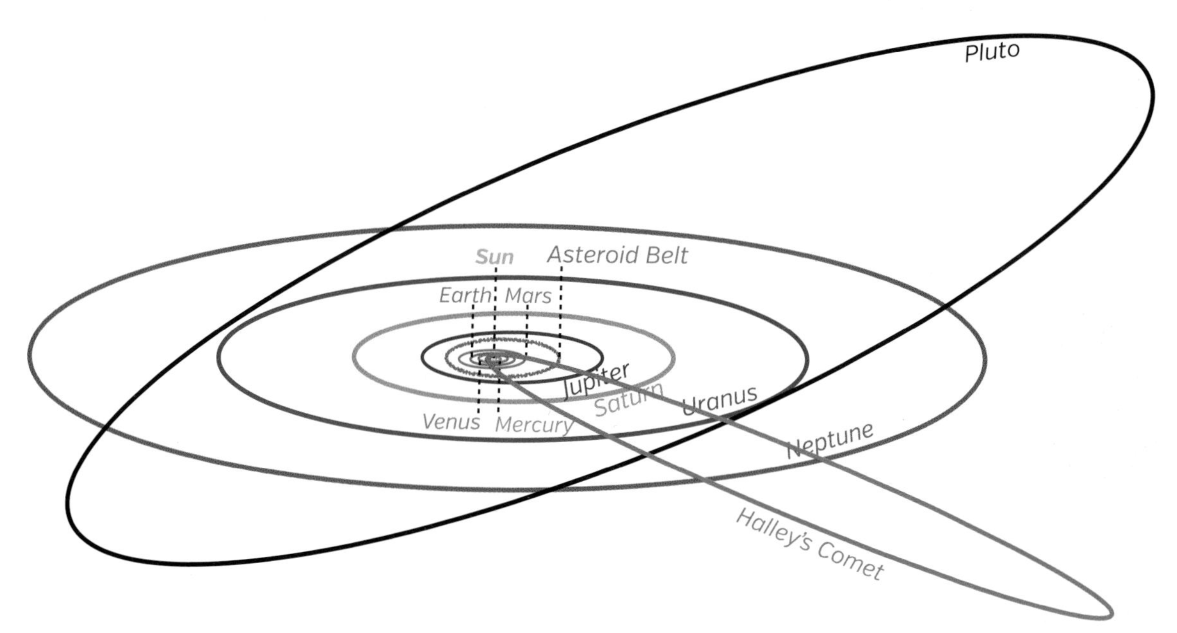

1 The probability of finding IMBH's at the needed locations and times is small, but in the spirit of science fiction, since it is within the bounds of physical law, we can utilize them.

Fig. 7.6. A single orbit of a spacecraft or planet or star around a gigantic, fast-spinning black hole such as Gargantua. *[From a simulation by Steve Drasco.]*

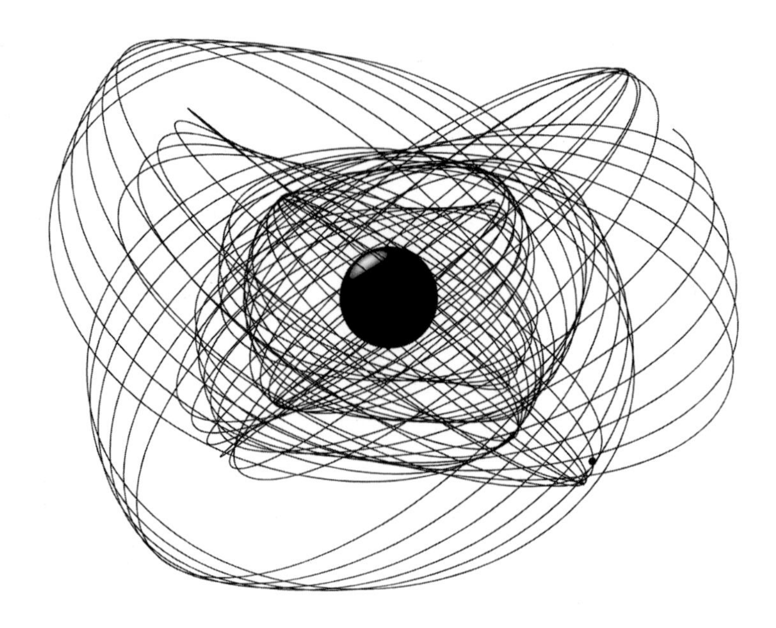

complex. Figure 7.6 is an example. For this orbit, each trip around Gargantua would require a few hours to a few days, so the entire pattern in Figure 7.6 would be swept out in about a year. After a few years, the orbit would pass near most any destination you might wish, though the speed at which you arrive might not be right. A slingshot might be needed to change speed and make a rendezvous.

I'll let *you* imagine how an ultra-advanced civilization might use such complex orbits. In my science interpretations of the movie, for simplicity I mostly eschew them and focus primarily on circular, equatorial orbits (those of the parked *Endurance*, Miller's planet, and the critical orbit), and on simple trajectories for the *Endurance* as it travels from one circular equatorial orbit to another. An exception is the orbit of Mann's planet, discussed in Chapter 19.

NASA's Gravitational Slingshots in the Solar System

Let's return from the world of the possible (what the laws of physics allow) to hard-nosed, real-life gravitational slingshots in the comfy confines of our solar system (what humans have actually achieved as of 2014).

You may be familiar with NASA's *Cassini* spacecraft (Figure 7.7). It was launched from Earth on October 15, 1997, with too little fuel to

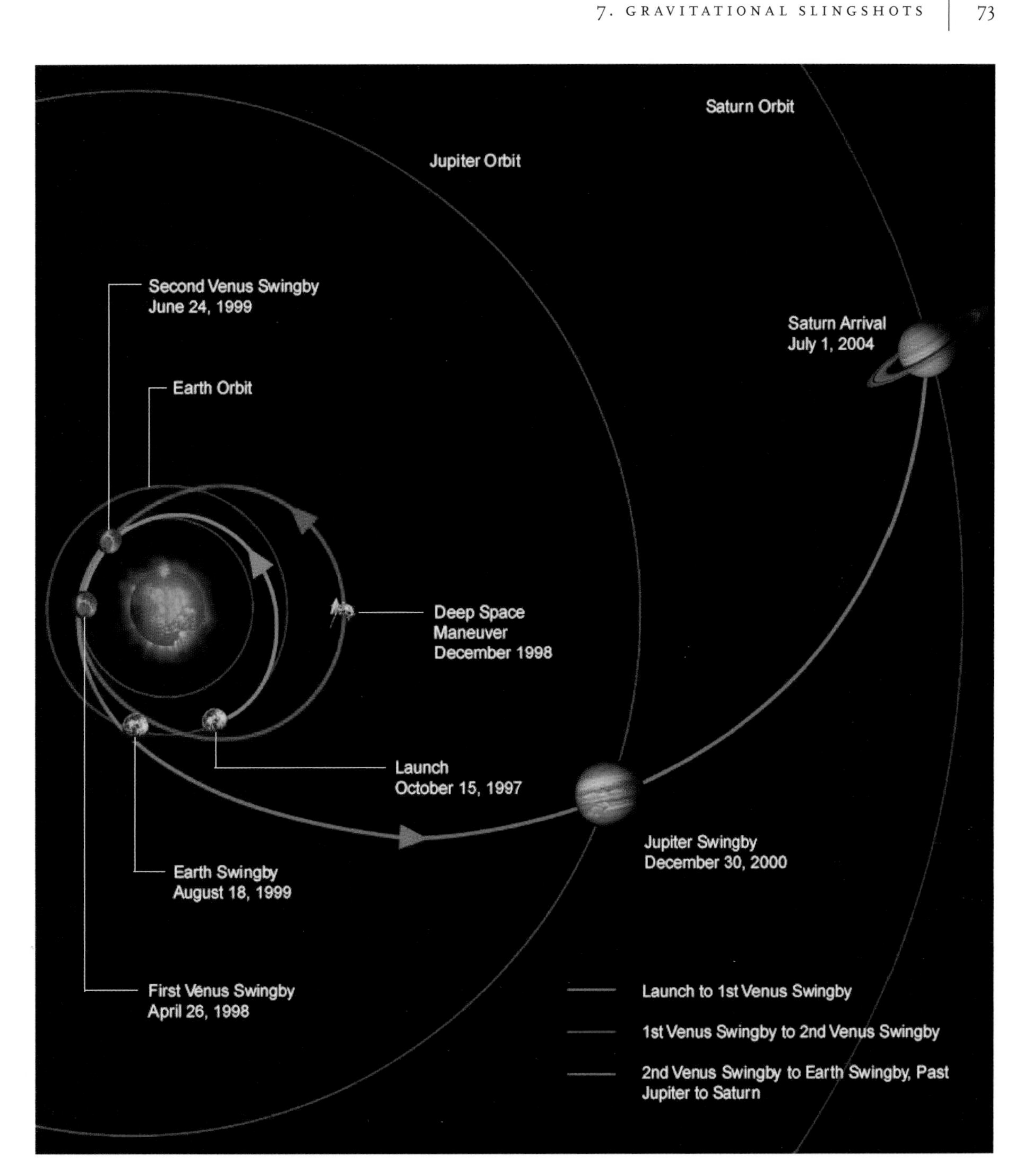

Fig. 7.7. The trajectory of *Cassini* from Earth to Saturn.

reach its destination, Saturn. The deficit was dealt with by slingshots: around Venus on April 26, 1998; a second slingshot around Venus on July 24, 1999; around Earth on August 18, 1999; and around Jupiter on December 30, 2000. Arriving at Saturn on July 1, 2004, *Cassini* slowed down with the aid of a slingshot around Saturn's closest moon, Io.

None of these slingshots looked like the ones I described above. Instead of strongly deflecting the spacecraft's direction of motion, Venus, Earth, Jupiter, and Io deflected it only mildly. Why?

The deflectors' gravity was too weak to produce a strong deflection. For Venus, Earth, and Io, the deflection was inevitably small because their gravity is intrinsically weak. Jupiter has much stronger gravity, but a large deflection would have sent *Cassini* in the wrong direction; reaching Saturn required a small deflection.

Despite the small deflections, *Cassini* got substantial kicks from the flybys, big enough to compensate for inadequate fuel. In each flyby (except Io), *Cassini* traveled behind the deflecting planet but at an angle, so the planet's gravity optimally pulled *Cassini* forward, speeding it up. In *Interstellar*, the *Endurance* does a similar slingshot around Mars.

Cassini has been exploring Saturn and Saturn's moons for the past ten years, sending back amazing images and information—a treasure trove of beauty and science. For a glimpse, see http://www.nasa.gov/mission_pages/cassini/main/.

By contrast with these weak slingshots in the solar system, Gargantua's intense gravity can grab even objects moving at ultrahigh speeds and throw them around on strongly bent slingshots. Even a light ray. This produces gravitational lensing, the key to seeing Gargantua.

8

Imaging Gargantua

Ⓣ

Black holes emit no light, so the only way to see Gargantua is by its influence on light from other objects. In *Interstellar* the other objects are an accretion disk (Chapter 9) and the galaxy in which it lives including nebulae and a rich field of stars. For the sake of simplicity, let's include only the stars for now.

Gargantua casts a black shadow on the field of stars and it also deflects the light rays from each star, distorting the stellar pattern that the camera sees. This distortion is the gravitational lensing discussed in Chapter 3.

Figure 8.1 shows a rapidly spinning black hole (let's call it Gargantua) in front of a field of stars, as it would appear to you if you were in Gargantua's equatorial plane. Gargantua's shadow is the totally black region. Immediately outside the shadow's edge is a very thin ring of starlight called the "ring of fire" that I intensified by hand to make the edge of the shadow more distinct. Outside that ring

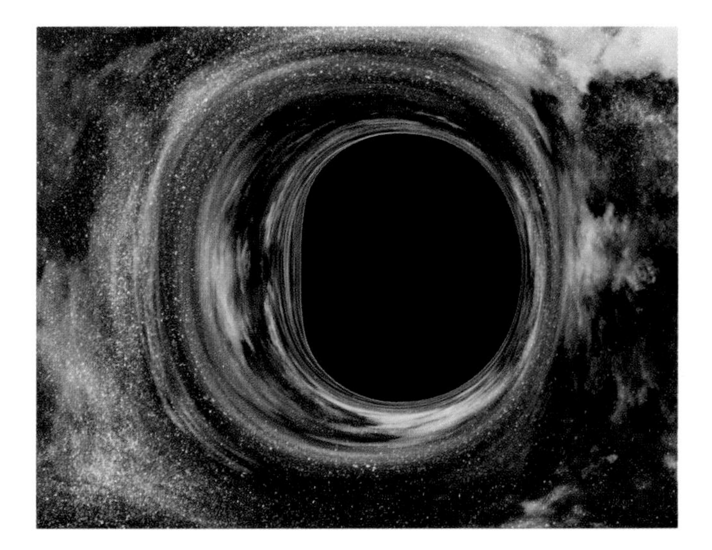

Fig. 8.1. The gravitationally lensed pattern of stars around a rapidly spinning black hole such as Gargantua. When seen from far away, the shadow's angular diameter, measured in radians, is 9 Gargantua radii divided by the observer's distance from Gargantua. *[From a simulation by the Double Negative visual-effects team.]*

we see a dense sprinkling of stars with a pattern of concentric shells, a pattern produced by the gravitational lensing.

As the camera orbits around Gargantua, the field of stars appears to move. This motion combined with the lensing produces dramatically changing patterns of light. The stars stream at high speed in some regions, they float gently in others, and they're frozen in still other regions; see the film clip on this book's page at Interstellar.withgoogle.com.

In this chapter I explain all these features, beginning with the shadow and its ring of fire. Then I describe how the black-hole images in *Interstellar* were actually produced.

When imaging Gargantua in this chapter, I treat it as a fast-spinning black hole, as it must be to produce the extreme loss of time that the *Endurance*'s crew experience relative to Earth (Chapter 6). However, for fast spin, a mass audience could be confused by the flattening of the left edge of Gargantua's shadow (Figure 8.1) and by some peculiar features of the star streaming and the accretion disk, so Christopher Nolan and Paul Franklin chose a smaller spin, 60 percent of the maximum, for their Gargantua images in the movie. See the last section in Chapter 9.

Warning: The explanations in the following three sections may require a lot of thought; you can skip them without losing pace with the rest of the book. Not to worry!

The Shadow and Its Ring of Fire

The shell of fire (Chapter 6) plays a key role in producing Gargantua's shadow and the thin ring of fire alongside it. The shell of fire is the purple region surrounding Gargantua in Figure 8.2, and it contains nearly trapped photon orbits (light rays) such as the one in the upper right inset.[1]

Suppose you are at the location of the yellow dot. The white light rays $\mathcal{A}$ and $\mathcal{B}$ and others like them bring you the image of the ring of fire, and the black light rays $\mathcal{A}$ and $\mathcal{B}$ bring you the image of the shadow's edge. For example, the white ray $\mathcal{A}$ originates at some star

1 See Figures 6.4 and 6.5.

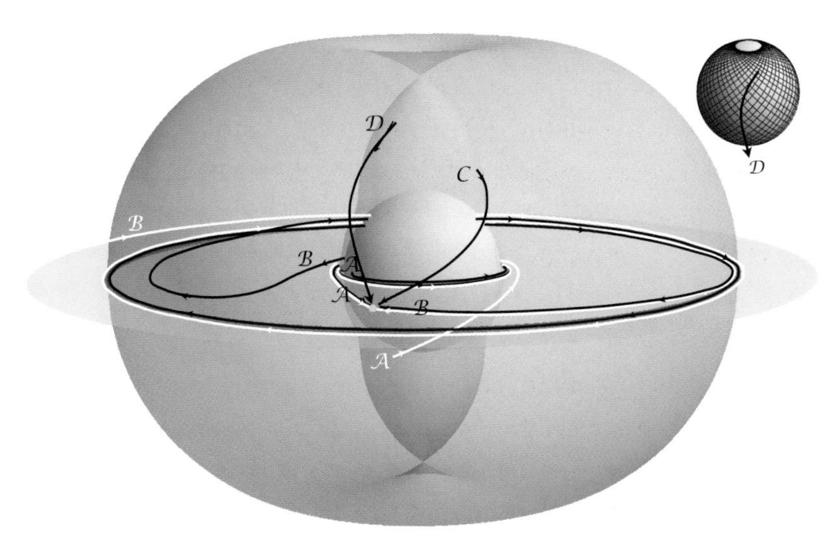

Fig. 8.2. Gargantua (*central spheroid*), its equatorial plane (blue), its shell of fire (*purple and violet*), and black and white light rays that bring you images of the shadow's edge and the thin ring alongside it.

far from Gargantua, it travels inward and gets trapped on the inner edge of the shell of fire in Gargantua's equatorial plane, where it flies round and round, driven by the whirl of space, and then escapes and comes to your eyes. The black ray also labeled 𝒜 originates on Gargantua's event horizon, it travels outward and gets trapped on that same inner edge of the shell of fire, where it goes round and round, then escapes and reaches your eyes alongside the white ray 𝒜. The white ray brings you an image of a bit of the thin ring; the black ray, an image of a bit of the shadow's edge. The shell of fire is responsible for merging the rays side by side and directing them toward your eyes.

Similarly for the white and black rays ℬ, except they get trapped on the outer edge of the shell of fire going clockwise (struggling against the whirl of space), while rays 𝒜 are trapped on the inner edge going counterclockwise (and driven by the whirl of space). The flattening of the shadow's left edge in Figure 8.1 and rounding of its right edge are due to rays 𝒜 (left edge) coming from the inner edge of the shell of fire, very close to the horizon, and rays ℬ (right edge) from the outer edge of the shell of fire, much further out.

Black rays 𝒞 and 𝒟 in Figure 8.2 begin on the horizon, travel outward and get trapped on nonequatorial orbits in the shell of fire, and then escape from their trapped orbits and come to your eyes, bringing images of bits of the shadow edge that lie outside the equatorial plane. The trapped orbit for ray 𝒟 is shown in the upper right inset. White rays *C* and *D* (not shown), coming from distant stars, get trapped alongside black rays *C* and *D*, and then travel to your eyes alongside *C* and *D*, bringing images of bits of the ring of fire alongside bits of the shadow edge.

Lensing by a Nonspinning Black Hole

To understand the pattern of gravitationally lensed stars outside the shadow and their streaming as the camera moves, let's begin with a nonspinning black hole and with light rays that emerge from a single star (Figure 8.3). Two light rays travel from the star to the camera. They each travel along the straightest line they can in the hole's warped space, but because of the warping, each ray gets bent.

One bent ray travels to the camera around the hole's left side; the other, around its right side. Each ray brings the camera its own image of the star. The two images, as seen by the camera, are shown in the inset of Figure 8.3. I put red circles around them to distinguish them from all the other stars the camera sees. Note that the right image is

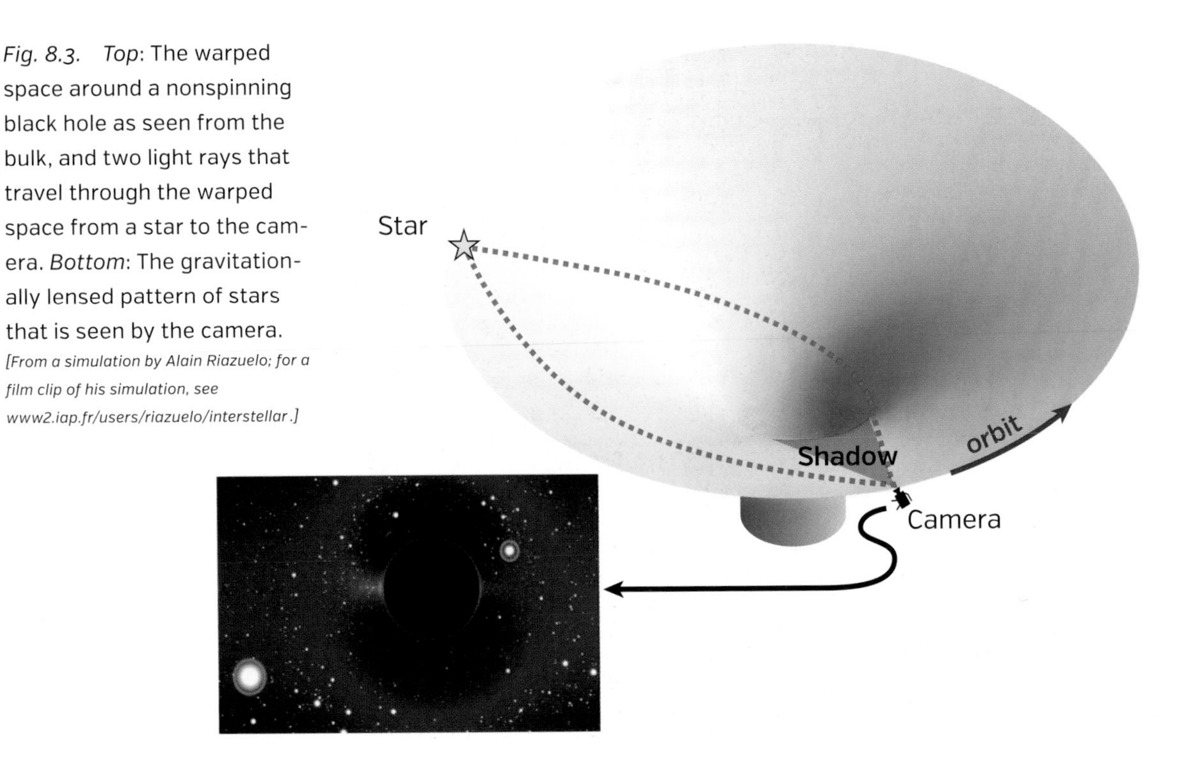

Fig. 8.3. *Top*: The warped space around a nonspinning black hole as seen from the bulk, and two light rays that travel through the warped space from a star to the camera. *Bottom*: The gravitationally lensed pattern of stars that is seen by the camera. *[From a simulation by Alain Riazuelo; for a film clip of his simulation, see www2.iap.fr/users/riazuelo/interstellar.]*

much closer to the hole's shadow than the left image. This is because its bent ray passed closer to the hole's event horizon.

Each of the other stars appears twice in the picture, on opposite sides of the hole's shadow. Can you identify some of the pairs? The black hole's shadow, in the picture, consists of directions from which no rays can come to the camera; see the triangular shaped region labeled "shadow" in the upper diagram. All the rays that "want to be" in the shadow got caught and swallowed by the black hole.

As the camera moves rightward in its orbit (Figure 8.3), the pattern of stars seen by the camera changes as shown in Figure 8.4.

This figure highlights two particular stars. One is circled in red (the same star circled in Figure 8.3). The other is inside a yellow diamond. We see two images of each star: one image is outside the pink circle; the other is inside the pink circle. This pink circle is called the "Einstein ring."

As the camera moves rightward, the images move along the yellow and red curves.

The star images outside the Einstein ring (the primary images,

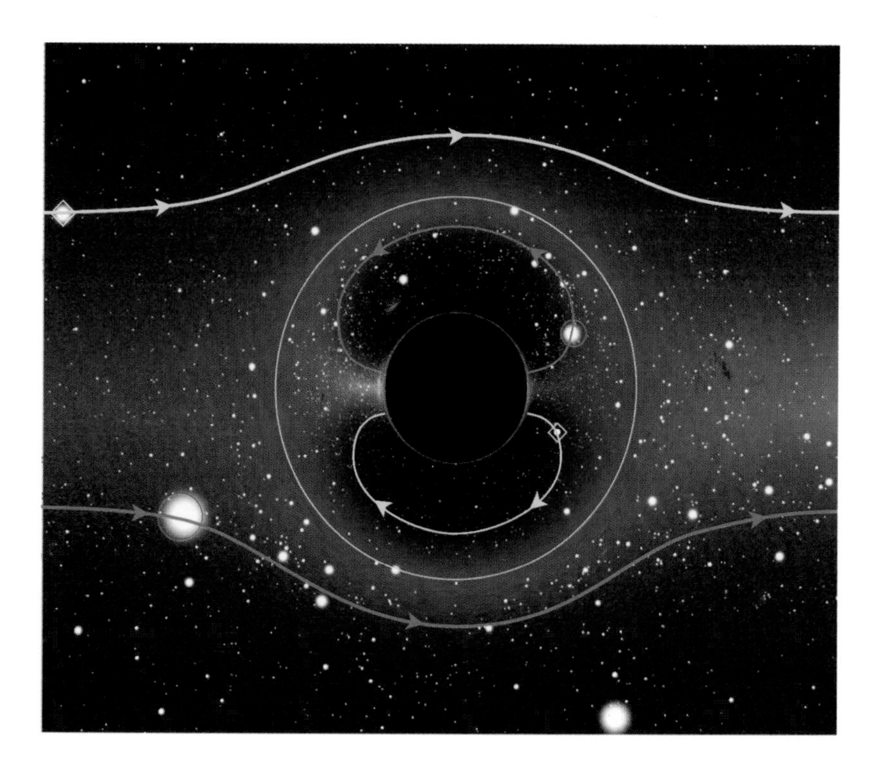

Fig. 8.4. The changing star pattern seen by the camera as it moves rightward in its orbit in Figure 8.3. *[From the simulation by Alain Riazuelo; see www2.iap.fr/users/riazuelo/interstellar.]*

let's call them) move in the way one might expect: smoothy from left to right, but deflecting away from the black hole as they move. (Can you figure out why the deflection is *away* from the hole instead of toward it?)

However the secondary images, inside the Einstein ring, move in an unexpected manner: They appear to emerge from the right edge of the shadow, move outward into the annulus between the shadow and the Einstein ring, swing leftward around the shadow, and descend back toward the shadow's edge.

You can understand this by going back to the upper drawing in Figure 8.3. The right ray passes near the black hole, so the right stellar image is near its shadow. Earlier in time, when the camera was further leftward, the right ray had to pass even closer to the black hole in order to bend more strongly and reach the camera, so the right image was very close to the edge of the shadow. By contrast, earlier in time, the left ray passed rather far from the hole and so was nearly straight and produced an image rather far from the hole.

Now, if you're ready, think through the subsequent motions of the images, depicted in Figure 8.4.

Lensing by a Rapidly Spinning Black Hole: Gargantua

The whirl of space generated by Gargantua's very fast spin changes the gravitational lensing. The star patterns in Figure 8.1 (Gargantua) look somewhat different from those in Figure 8.4 (a nonspinning black hole), and the streaming patterns differ even more.

For Gargantua the streaming (Figure 8.5) reveals two Einstein rings, shown as pink curves. Outside the outer ring, the stars stream rightward (for example, along the two red curves), as they did for a nonspinning black hole in Figure 8.4. However, the whirl of space has concentrated the stream into narrowed high-speed strips along the back edge of the hole's shadow, strips that bend somewhat sharply at the equator. The whirl has also produced eddies in the streaming (the closed red curves).

The secondary image of each star appears between the two Einstein rings. Each secondary image circulates along a closed curve (for

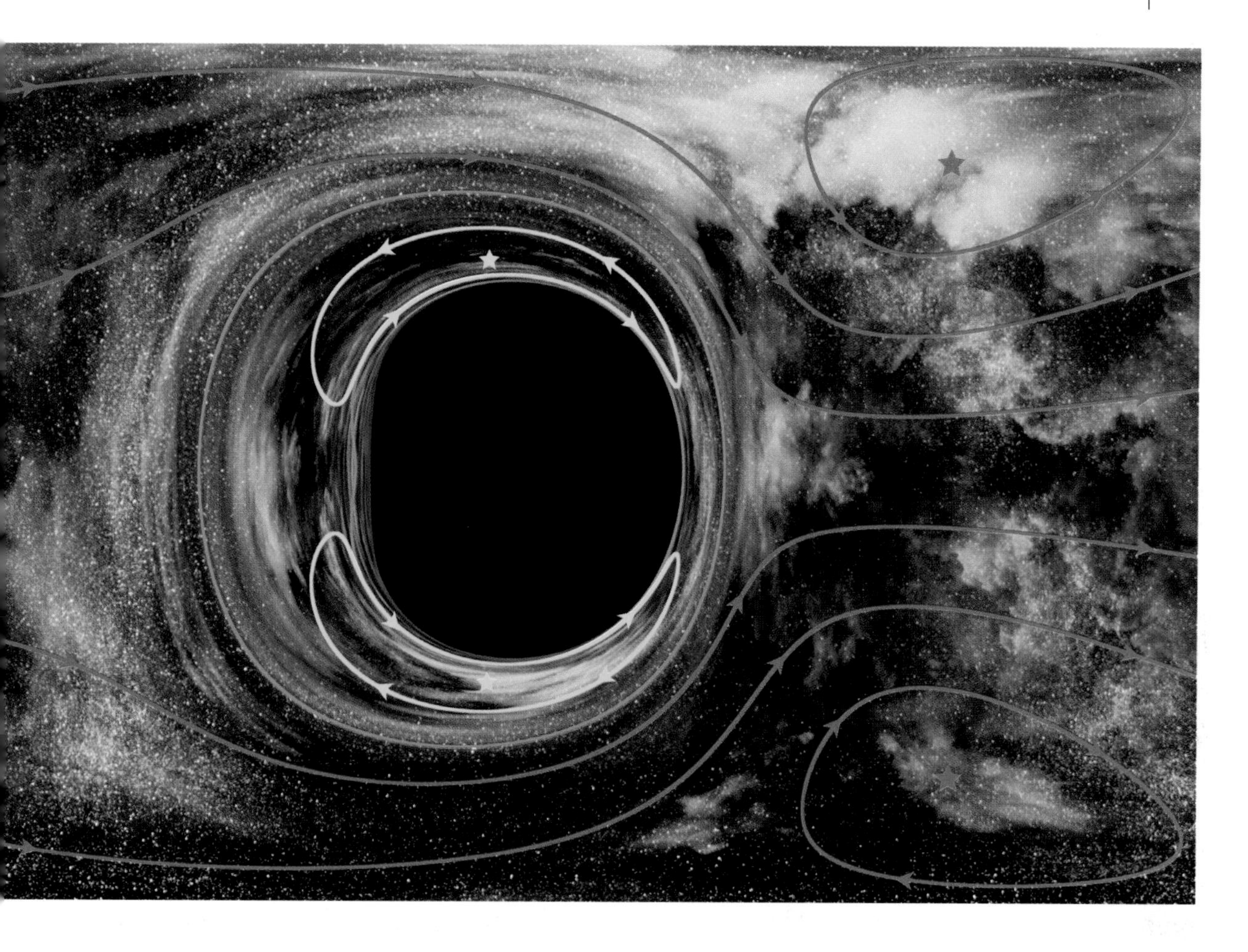

example, the two yellow curves), and it circulates in the opposite direction to the red streaming motions outside the outer ring.

There are two very special stars in Gargantua's sky with gravitational lensing turned off. One lies directly above Gargantua's north pole; the other directly below its south pole. These are analogs of the star Polaris, which resides directly above the Earth's north pole. I placed five-pointed stars at the primary (red) and secondary (yellow) images of Gargantua's pole stars. All the stars on the Earth's sky appear to circulate around Polaris as we humans are carried around by the Earth's rotation. Similarly, all of Gargantua's primary stellar images circulate around the red pole-star images as the camera orbits the hole, but their circulation paths (for example, the two red eddy curves) are highly distorted by the whirl of space and gravitational lensing. Similarly, all the secondary stellar images circulate around

Fig. 8.5. The star streaming patterns as seen by a camera near a rapidly spinning black hole such as Gargantua. In this simulation by the Double Negative visual-effects team, the hole spins at 99.9 percent of the fastest possible, and the camera is in a circular, equatorial orbit with circumference six times larger than the horizon's circumference. For a film clip of this simulation, see this book's page at Interstellar.withgoogle.com.

the yellow pole-star images (for example, along the two distorted yellow curves).

Why, for a nonspinning hole (Figure 8.4), did the secondary images appear to emerge from the black hole's shadow, swing around the hole, and descend back into the shadow, instead of circulating around a closed curve as for Gargantua (Figure 8.5)? They actually *do* circulate around closed curves for a nonspinning hole. However, the inner edge of the closed curve is so close to the shadow's edge that it can't be seen. Gargantua's spin makes space whirl, and that whirl moves the inner Einstein ring outward, revealing the secondary images' full circulatory pattern (yellow curves in Figure 8.5), and revealing the inner Einstein ring.

Inside the inner Einstein ring, the streaming pattern is more complicated. The stars in this region are tertiary and higher-order images of all the stars in the universe—the same stars as appear as primary images outside the outer Einstein ring and secondary images between the Einstein rings.

In Figure 8.6, I show five small pictures of Gargantua's equatorial plane, with Gargantua itself in black, the camera's orbit in dashed purple, and a light ray in red. The light ray brings to the camera the stellar image that is at the tip of the blue arrow. The camera is moving counterclockwise around Gargantua.

You can get a lot of insight into the gravitational lensing by walking yourself through these pictures, one by one. Take note: The actual direction to the star is upward and rightward (see outer ends of the red rays). The camera and beginning of each ray point toward the stellar image. The tenth image is very near the left edge of the shadow and the right secondary image is near the right edge; comparing the directions of the camera for these images, we see that the shadow subtends about 150 degrees in the upward direction. This despite the fact that the actual direction from camera to center of Gargantua is leftward and upward. The lensing has moved the shadow relative to Gargantua's actual direction.

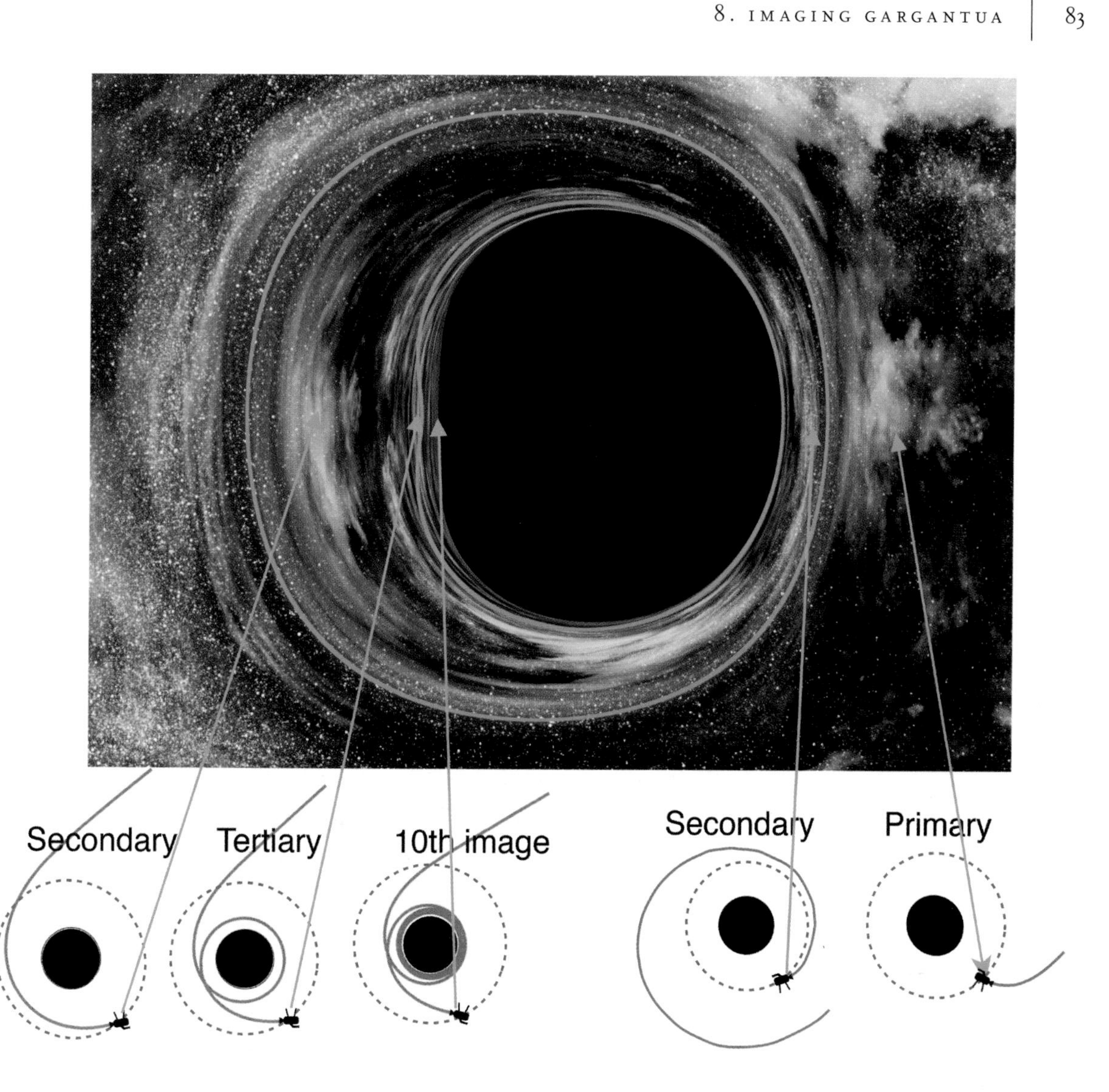

Fig. 8.6. Light rays that bring images of the stars at the tips of the blue arrows. *[From the same Double Negative simulation as Figures 8.1 and 8.5.]*

Creating *Interstellar*'s Black-Hole and Wormhole Visual Effects

Chris wanted Gargantua to look like what a spinning black hole *really* looks like when viewed up close, so he asked Paul to consult with me. Paul put me in touch with the *Interstellar* team he had assembled at his London-based visual-effects studio, Double Negative.

I wound up working closely with Oliver James, the chief scientist. Oliver and I talked by phone and Skype, exchanged e-mails and electronic files, and met in person in Los Angeles and at his London office. Oliver has a college degree in optics and atomic physics and understands Einstein's relativity laws, so we speak the same technical language.

Several of my physicist colleagues had already done computer simulations of what one would see when orbiting a black hole and even falling into one. The best experts were Alain Riazuelo, at the Institut d'Astrophysique in Paris, and Andrew Hamilton, at the University of Colorado in Boulder. Andrew had generated black-hole movies shown in planetariums around the world, and Alain had simulated black holes that spin very, very fast, like Gargantua.

So initially I planned to put Oliver in touch with Alain and Andrew and ask them to provide him the input he needed. I lived uncomfortably with that decision for several days, and then changed my mind.

During my half century physics career I put great effort into making new discoveries myself and mentoring students as they made new discoveries. Why not, for a change, do something just because it's fun, I asked myself, even though others have done it before me? And so I went for it. And it *was* fun. And to my surprise, as a byproduct, it produced (modest) new discoveries.

Using Einstein's relativistic laws of physics and leaning heavily on prior work by others (especially Brandon Carter at the Laboratoire Univers et Théories in France and Janna Levin at Columbia University), I worked out the equations Oliver needed. These equations compute the trajectories of light rays that begin at some light source, for example, a distant star, and travel inward through Gargantua's warped space and time to the camera. From those light rays, my equations then compute the images the camera sees, taking account not only of the light's sources and Gargantua's warping of space and time, but also the camera's motion around Gargantua.

Having derived the equations, I implemented them myself, using user-friendly computer software called Mathematica. I compared images produced by my Mathematica computer code with Alain Riazuelo's images, and when they agreed, I cheered. I then wrote up detailed descriptions of my equations and sent them to Oliver in London, along with my Mathematica code.

My code was very slow and had low resolution. Oliver's challenge was to convert my equations into computer code that could generate the ultra-high-quality IMAX images needed for the movie.

Oliver and I did this in steps. We began with a nonspinning black hole and a nonmoving camera. Then we added the black hole's spin. Then we added the camera's motion: first motion in a circular orbit, and then plunging into a black hole. And then we switched to a camera around a wormhole.

At this point, Oliver hit me with a minibombshell: To model some of the more subtle effects, he would need not only equations describing the trajectory of a ray of light, but also equations describing how the cross section of a beam of light changes its size and shape during its journey past the black hole.

I knew more or less how to do this, but the equations were horrendously complicated and I feared making mistakes. So I searched the technical literature and found that in 1977 Serge Pineault and Rob Roeder at the University of Toronto had derived the necessary equations in almost the form I needed. After a three-week struggle with my own stupidities, I brought their equations into precisely the needed form, implemented them in Mathematica, and wrote them up for Oliver, who incorporated them into his own computer code. At last his code could produce the quality images needed for the movie.

At Double Negative, Oliver's computer code was just the beginning. He handed it over to an artistic team led by Eugénie von Tunzelmann, who added an accretion disk (Chapter 9) and created the background galaxy with its stars and nebulae that Gargantua would lens. Her team then added the *Endurance* and Rangers and landers and the camera animation (its changing motion, direction, field of view, etc.), and molded the images into intensely compelling forms: the fabulous scenes that actually appear in the movie. For further discussion, see Chapter 9.

In the meantime, I puzzled over the high-resolution film clips that Oliver and Eugénie sent me, struggling to extract insights into why the images look like they look, and why the star fields stream as they stream. For me, those film clips are like experimental data: they reveal things I never could have figured out on my own, without those simulations—for example, the things I described in the previous sec-

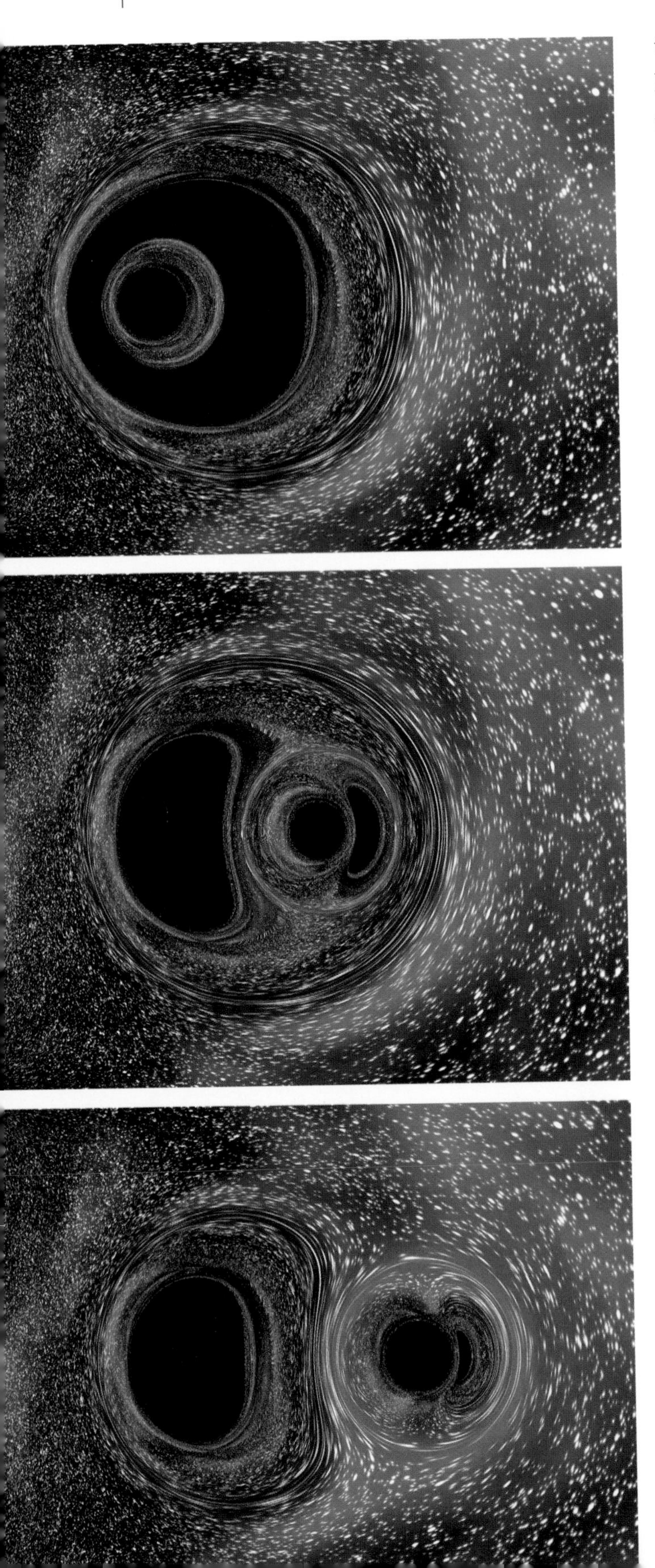

tion (Figures 8.5 and 8.6). We plan to publish one or more technical papers, describing the new things we learned.

Imaging a Gravitational Slingshot

Although Chris chose not to show any gravitational slingshots in *Interstellar*, I wondered what they would look like to Cooper as he piloted the Ranger toward Miller's planet. So I used my equations and Mathematica to simulate them and produce images. (My images have far lower resolution than Oliver's and Eugénie's due to my code's slowness.)

Figure 8.7 shows a sequence of images, as Cooper's Ranger swings around an intermediate-mass black hole (IMBH) to initiate its descent toward Miller's planet—in my scientist's interpretation of *Interstellar*. This is the slingshot described in Figure 7.2.

In the top image, Gargantua is in the background with the IMBH passing in front of it. The IMBH grabs light rays from distant stars that are headed toward gargantua, swings the rays around itself, and ejects them toward the camera. This explains the donut of starlight that surrounds the IMBH's shadow. Although the IMBH is a thousand times smaller than Gargantua, it is far closer to the Ranger than is Gargantua, so it looks only modestly smaller.

As the IMBH appears to move rightward, as seen by the slingshot-moving cam-

Fig. 8.7. Gravitational slingshot around an IMBH, with Gargantua in the background. *[My own simulation and visualization.]*

era, it leaves Gargantua's primary shadow behind itself (middle picture in Figure 8.7), and it pushes a secondary image of Gargantua's shadow ahead of itself. These two images are completely analogous to the primary and secondary images of a star gravitationally lensed by a black hole; but now it is Gargantua's shadow that is being lensed, by the IMBH. In the bottom picture, the secondary shadow is shrinking in size, as the IMBH moves onward. By this time the slingshot is nearly complete, and the camera, on board the Ranger, is headed downward, toward Miller's planet.

As impressive as these images may be, they can be seen only up close to the IMBH and Gargantua, not from the great distance of Earth. To astronomers on Earth, the most visually impressive things about gigantic black holes are jets that stick out of them and the light from brilliant disks of hot gas that orbit them. To these we'll now turn.

9

Disks and Jets

Ⓣ

Quasars

Most of the objects seen by radio telescopes are huge clouds of gas, clouds far larger than any star. But in the early 1960s a few tiny objects were found. Astronomers named these objects *quasars* for "quasi-stellar radio sources."

In 1962 the Caltech astronomer Maarten Schmidt, looking through the world's largest optical telescope on Palomar Mountain, discovered light coming from a quasar called 3C273. It looked like a bright star with a faint jet shooting out of it (Figure 9.1). This was weird!

When Schmidt split 3C273's light into its various colors (as is sometimes done by sending light through a prism), he saw the set of spectral lines in the bottom of Figure 9.1. At first sight, these were unlike any spectral lines he had ever seen. But in February 1963, after a few months' struggle, he realized the lines were unfamiliar simply because their wavelengths were 16 percent larger than normal. This is called the Doppler shift; it was caused by the quasar's moving away from Earth at 16 percent the speed of light, about $c/6$. What could cause that ultrafast motion? The least crazy explanation Schmidt could find was the expansion of the universe.

As the universe expands, objects far from Earth move apart from us at very high speed, and objects nearer move away more slowly. 3C273's enormous speed, one-sixth that of light, meant that 3C273 was 2 billion light-years from Earth, nearly the farthest object that had ever been seen at that time. From its brightness and its distance, Schmidt concluded that 3C273 puts out 4 trillion times more power in light than the Sun, and a hundred times more power than the brightest galaxies!

This prodigious power fluctuated on times as short as a month, so most of the light must be coming from an object so small that the light can travel across it in one month's time—far smaller than the distance from Earth to the nearest star, Proxima Centauri. And other quasars with almost as much power fluctuated on times of a few hours, so they had to be not much larger than our solar system. *One hundred times the power of a bright galaxy, coming from a region the size of our solar system; that was phenomenal!*

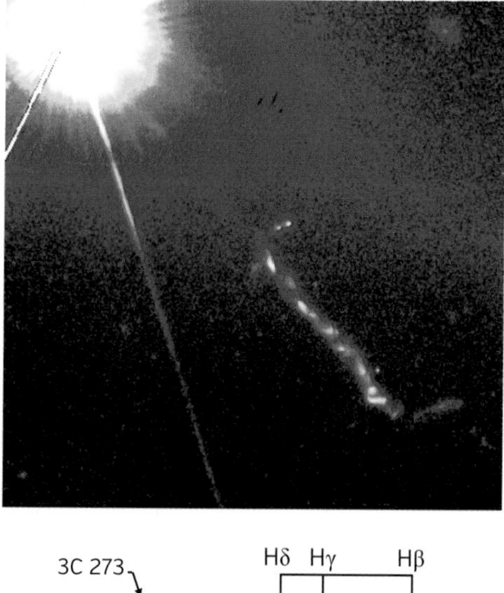

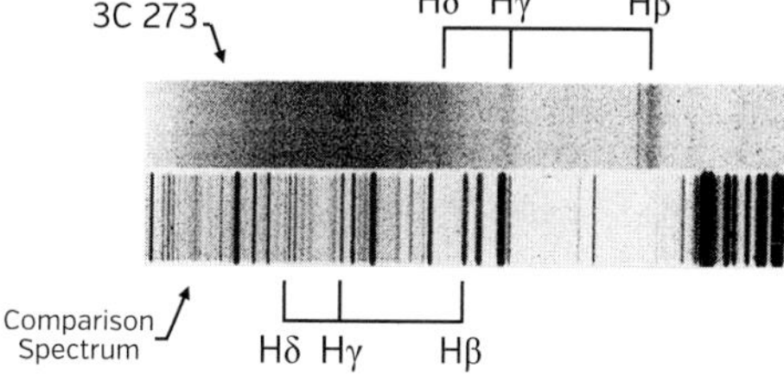

Fig. 9.1. *Top*: Photograph of 3C273 taken by NASA's Hubble Space Telescope. The star (*upper left*) looks big only because the photo is overexposed in order to see the faint jet (*lower right*). It is actually so small that its size cannot be measured. *Bottom*: Maarten Schmidt's spectral lines from 3C273 (*upper panel*) compared with spectral lines of hydrogen measured in an Earth laboratory. The quasar's three lines are the same as hydrogen's lines called Hβ, Hγ, and Hδ, but with wavelengths increased by 16 percent. (The images of the spectral lines are photographic negatives: black lines are really bright.)

Black Holes and Accretion Disks

How could so much power come out of a region so small? When we think about the fundamental forces in Nature, there are three possibilities: chemical energy, nuclear energy, or gravitational energy.

Chemical energy is the energy released when molecules combine together to make new kinds of molecules. An example is burning gasoline, which combines oxygen from the air with gasoline molecules to make water and carbon dioxide, and a lot of heat. The power from that would be far, far, far too little though.

Nuclear energy results when atomic nuclei combine together to make new atomic nuclei. Examples are an atomic bomb, a hydrogen bomb, and the burning of nuclear fuel inside a star. Though this can be far more powerful than chemical energy (think of the difference between a gasoline fire and a nuclear bomb), astrophysicists couldn't see any plausible way for nuclear energy to power quasars. It was still too puny.

So the only possibility left was *gravitational energy*, the same kind of energy we were driven to, when navigating the *Endurance* around Gargantua. For the *Endurance*, gravitational energy was harnessed by a slingshot around an intermediate-mass black hole (Chapter 7). The black hole's intense gravity was key. For quasars, similarly, the power must come from a black hole.

For several years, astrophysicists struggled to figure out how a black hole could do the job. The answer was found in 1969, by Donald Lynden-Bell at the Royal Greenwich Observatory in England. A quasar, Lynden-Bell hypothesized, is a gigantic black hole surrounded by a disk of hot gas (an accretion disk) that is threaded by a magnetic field (Figure 9.2).

Hot gas in our universe is almost always threaded by magnetic fields (Chapter 2). These fields are locked into the gas; the gas and fields move together, in lockstep.

When threading an accretion disk, a magnetic field becomes a catalyst for converting gravitational energy into heat and then light. The field provides ultrastrong friction[1] that slows the gas's circumferential motion, reducing the centrifugal force that holds it out against the pull of gravity, so the gas moves inward, toward the black hole. As the gas moves inward, the hole's gravity speeds up its orbital motion by even more than the friction slowed it. In other words, gravitational energy is converted into kinetic energy (energy of motion). Magnetic friction then converts half that new kinetic energy into heat and light, and the process repeats.

The energy comes from the black hole's gravity. The agents for extracting it are magnetic friction and the disk's gas.

The quasar's bright light, seen by astronomers, comes from the disk's heated gas, Lynden-Bell concluded. Moreover, the magnetic field accelerates some of the gas's electrons to high energies; and the electrons then spiral around the magnetic force lines, emitting the quasar's observed radio waves.

1 The friction arises through a complex process where moving gas winds the field up, strengthening it and thereby converting energy of motion into magnetic energy; and then the magnetic field, pointing in opposite directions in neighboring regions of space, reconnects and in the process converts magnetic energy into heat. That's the nature of friction: a conversion of motion into heat.

Lynden-Bell worked out the details of all this using a combination of the Newtonian, relativistic, and quantum laws of physics. He easily explained everything about quasars that astronomers had seen, except their jets. His technical article describing his reasoning and his calculations (Lynden-Bell 1979) is one of the great astrophysics articles of all time.

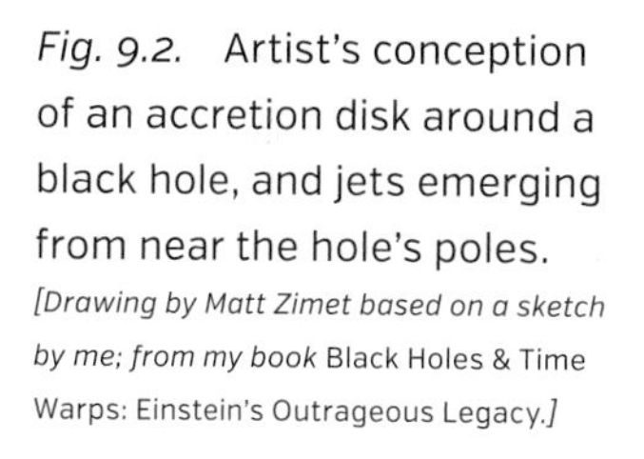

Fig. 9.2. Artist's conception of an accretion disk around a black hole, and jets emerging from near the hole's poles. *[Drawing by Matt Zimet based on a sketch by me; from my book* Black Holes & Time Warps: Einstein's Outrageous Legacy.*]*

The Jets: Extracting Power from Whirling Space

Over the next few years, astronomers discovered many more jets sticking out of quasars and studied them in great detail. It soon became clear that they are streams of hot, magnetized gas ejected from the quasar itself: from the black hole and its accretion disk (Figure 9.2). And the ejection is extremely powerful: the gas travels out the jets at nearly the speed of light. As it travels, and when it plows into material far from the quasar, the gas emits power in light, in radio waves, in X-rays, and even in gamma rays. The jets are sometimes as bright as the quasar itself, a hundred times brighter than the brightest galaxies.

Astrophysicists struggled for nearly a decade to explain how the jets are powered and what makes them so fast, so narrow, and so straight. The answers came in several variants, with the most interesting in 1977 from Roger Blandford at the University of Cambridge, England, and his student Roman Znajek, building on foundations laid by the Oxford physicist Roger Penrose; see Figure 9.3.

The accretion disk's gas gradually spirals into the black hole. When crossing the hole's event horizon, each bit of gas deposits its bit of magnetic field onto the horizon, and then the surrounding disk holds it there, Blandford and Znajek concluded. As the black hole spins, it drags space into whirling motion (Figures 5.4 and 5.5), and the whirl-

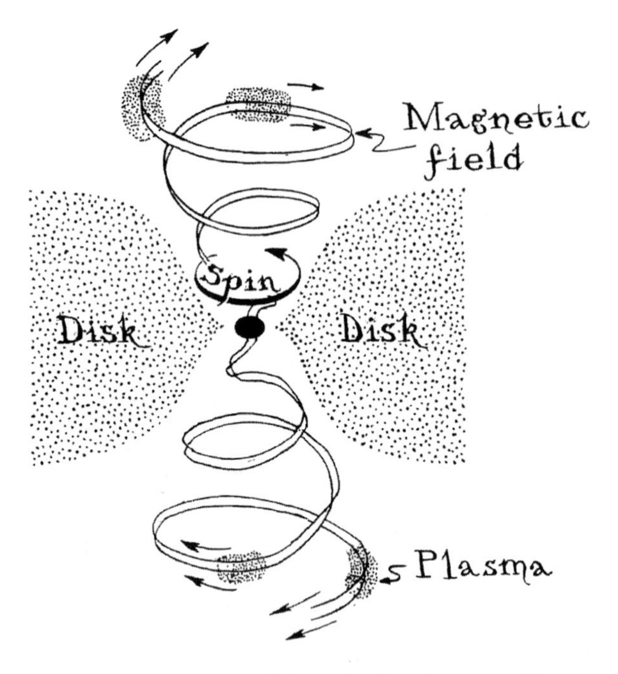

Fig. 9.3. Blandford-Znajek mechanism for generating jets. *[Drawing by Matt Zimet based on a sketch by me; from my book* Black Holes & Time Warps: Einstein's Outrageous Legacy.*]*

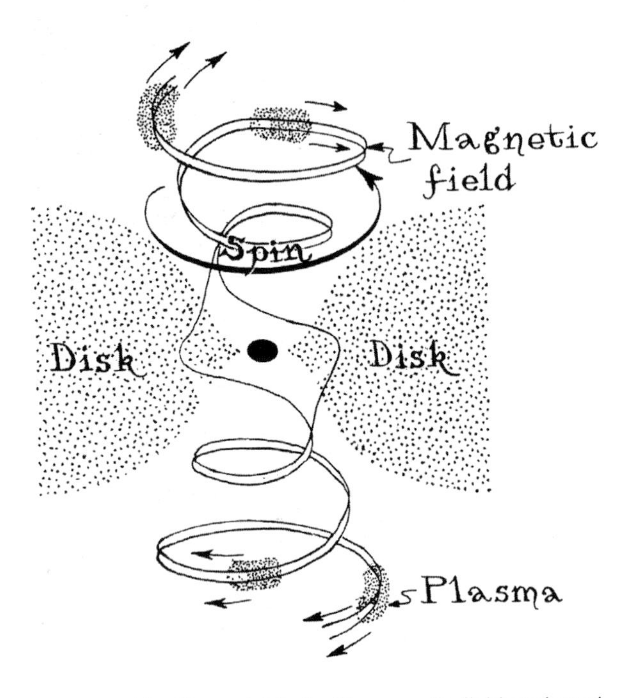

Fig. 9.4. Like Figure 9.3 but with magnetic field anchored in the accretion disk. *[Drawing by Matt Zimet based on a sketch by me; from my book* Black Holes & Time Warps: Einstein's Outrageous Legacy.*]*

ing space makes the magnetic field whirl (Figure 9.3). The whirling magnetic field generates an intense electric field like in a dynamo at a hydroelectric power station. The electric field and the whirling magnetic field together fling plasma (hot, ionized gas) upward and downward at near light speed, creating and powering two jets. The jets' directions are held steady (when averaged over years) by the black hole's spin, which is steady due to gyroscopic action.

In 3C273 only one jet was bright enough to see (Figure 9.1), but in many other quasars both are seen.

Blandford and Znajek worked out the full details, relying heavily on Einstein's relativistic laws. They were able to explain most everything about the jets that astronomers see.

In a second variant of the explanation (Figure 9.4), the whirling magnetic field is anchored in the accretion disk instead of the hole, and is dragged around by the disk's orbital motion. Otherwise, the story is the same: dynamo action; plasma flung out. This variant works well even if the black hole isn't spinning. But we're pretty sure that most black holes spin fast, so I suspect the Blandford-Znajek mechanism (Figure 9.3) is the most common one in quasars. However, I may be prejudiced. I spent much time in the 1980s exploring aspects of the Blandford-Znajek ideas and even coauthored a technical book about them.

Whence Comes the Disk? Tidal Forces Tear Stars Apart

Lynden-Bell, in 1969, speculated that quasars live at the centers of galaxies. We don't see a quasar's host galaxy, he said, because its light is so much fainter than the quasar's light. The quasar drowns the galaxy out. In the decades since then, with improving technology, astronomers have found the galaxy's light around many quasars, confirming Lynden-Bell's speculation.

In those recent decades we also learned where most of the disk's gas comes from. Occasionally a star strays so close to the quasar's black hole that the hole's tidal gravity (Chapter 4) tears the star apart. Much of the shredded star's gas is captured by the black hole and forms an accretion disk, but some of the gas escapes.

In recent years, thanks to improving computer technology, astrophysicists simulated this. Figure 9.5 is from a recent simulation by James Guillochon, Enrico Ramirez-Ruiz, and Daniel Kasen (University of California at Santa Cruz) and Stephan Rosswog (University of Bremen).[2] At time zero (not shown) the star was headed almost precisely toward the black hole and the hole's tidal gravity was beginning to stretch the star toward the hole and squeeze it from the sides, as in Figure 6.1. Twelve hours later the star is strongly deformed and at the location shown in Figure 9.5. Over the next few hours, it swings around the hole along the blue gravitational-slingshot orbit and deforms further as shown. By twenty-four hours the star is flying apart; its own gravity can no longer hold it together.

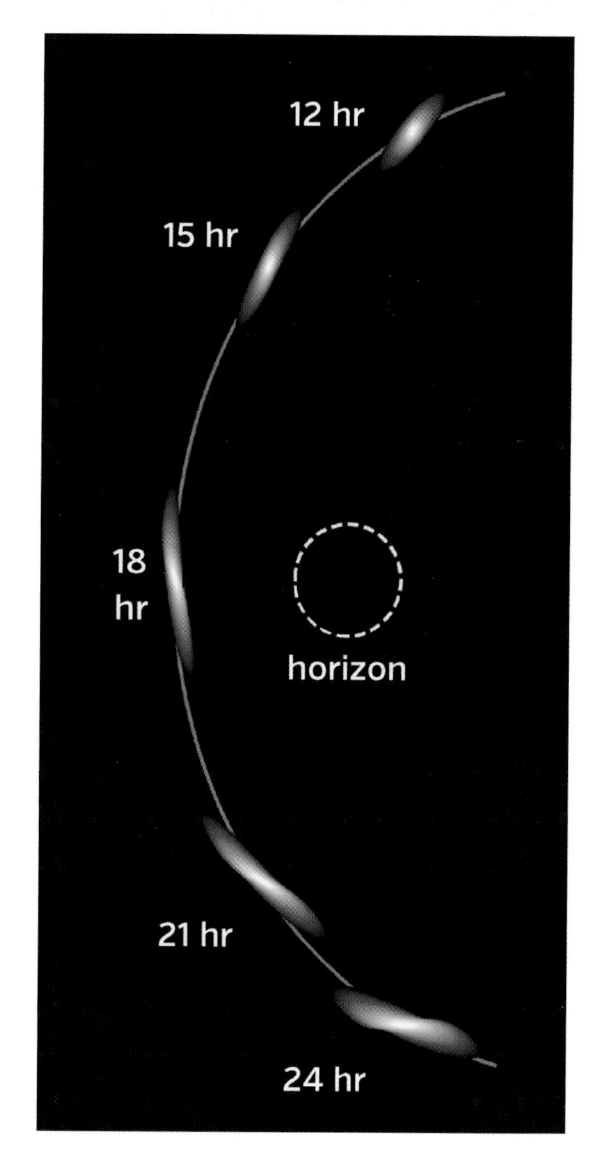

Fig. 9.5. Tidal disruption of a red giant star by a black hole similar to Gargantua.

2 I changed the size of the hole to that of Gargantua and the size of the star to that of a red giant, and changed the time markers in Figure 9.5 accordingly.

The star's subsequent fate is shown in Figure 9.6, from a different simulation by James Guillochon together with Suvi Gezari (Johns Hopkins University). For a movie of this simulation, see http://hubblesite.org/newscenter/archive/releases/2012/18/video/a/.

The top two images are shortly before the beginning and shortly after the end of Figure 9.5; I enlarged these two images tenfold compared to the others, to make the hole and the disrupting star visible.

As the whole set of images shows, over the subsequent several years much of the star's matter is captured into orbit around the black hole, where it begins to form an accretion disk. And the remaining matter escapes from the hole's gravitational pull along a streaming, jetlike trajectory.

Gargantua's Accretion Disk and Missing Jet

A typical accretion disk and its jet emit radiation—X-rays, gamma rays, radio waves, and light—radiation so intense that it would fry any human nearby. To avoid frying, Christopher Nolan and Paul Franklin gave Gargantua an exceedingly anemic disk.

Now, "anemic" doesn't mean anemic by human standards; just by the standards of typical quasars. Instead of being a hundred million degrees like a typical quasar's disk, Gargantua's disk is only a few thousand degrees, like the Sun's surface, so it emits lots of light but little to no X-rays or gamma rays. With gas so cool, the atoms' thermal motions are too slow to puff the disk up much. The disk is thin and nearly confined to Gargantua's equatorial plane, with only a little puffing.

Disks like this might be common around black holes that have not torn a star apart in the past millions of years or more—that have not been "fed" in a long time. The magnetic field, originally confined by the disk's plasma, may have largely leaked away. And the jet, previously powered by the magnetic field, may have died. Such is Gargantua's disk: jetless and thin and relatively safe for humans. Relatively.

Gargantua's disk looks quite different from the pictures of thin disks that you see on the web or in astrophysicists' technical publications, because those pictures omit a key feature: the gravitational lens-

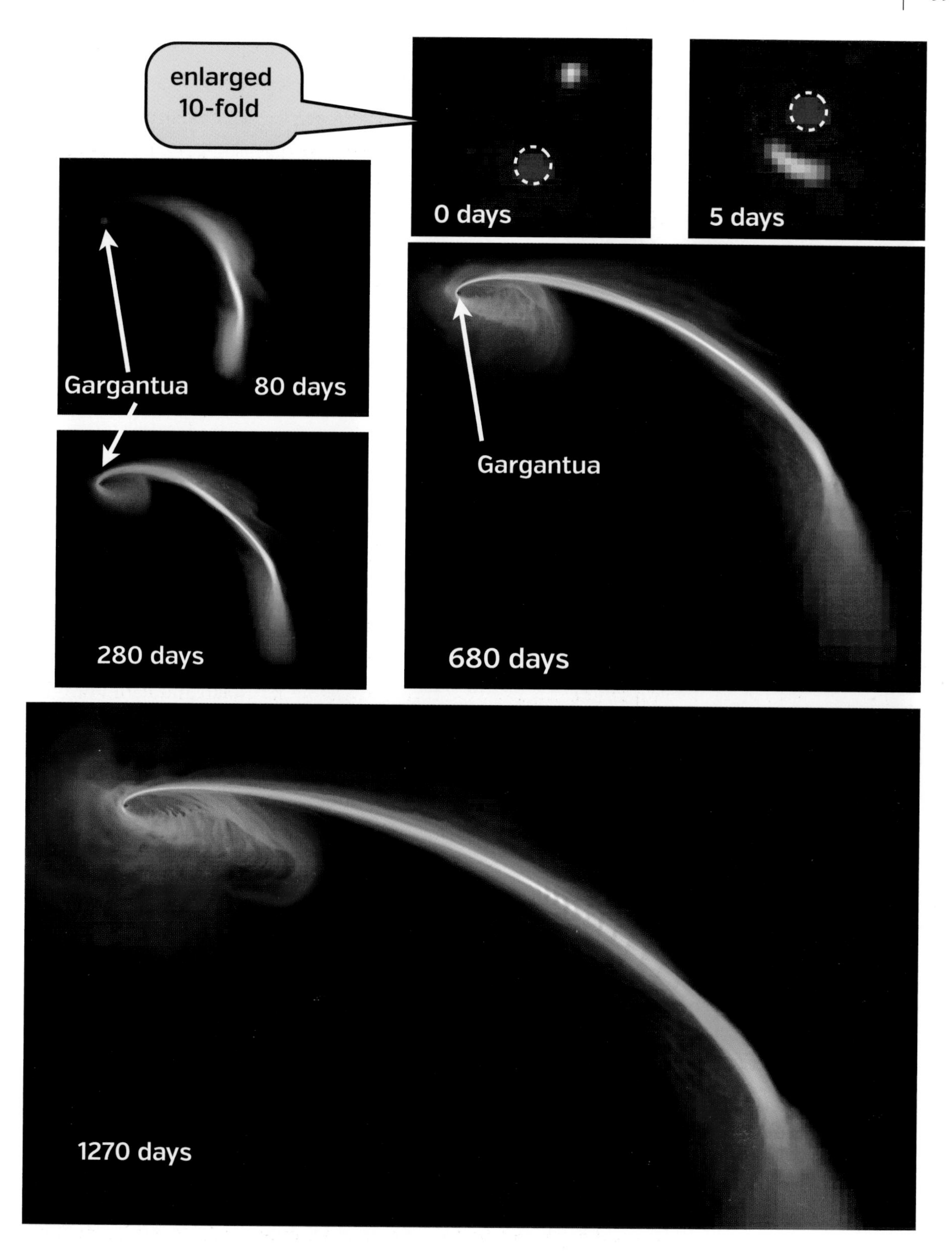

Fig. 9.6. Subsequent fate of the star in Figure 9.5.

ing of the disk by its black hole. Not so in *Interstellar,* where Chris insisted on visual accuracy.

Eugénie von Tunzelmann was charged with putting an accretion disk into Oliver James' gravitational lensing computer code, the code I described in Chapter 8. As a first step, just to see what the lensing does, Eugénie inserted a disk that was truly infinitesimally thin and lay precisely in Gargantua's equatorial plane. For this book she has provided a more pedagogical version of that disk, made of equally spaced color swatches (Inset in Figure 9.7).

If there had been no gravitational lensing, the disk would have looked like the inset. The lensing produced huge changes from this (body of Figure 9.7). You might have expected the back portion of the disk to be hidden behind the black hole. Not so. Instead, it is gravitationally lensed to produce two images, one above Gargantua and the other below; see Figure 9.8. Light rays emitted from the disk's top face, behind Gargantua, travel up and over the hole to the camera, producing the disk image that wraps over the top of Gargantua's shadow in Figure 9.7; and similarly for the disk image that wraps under the bottom of Gargantua's shadow.

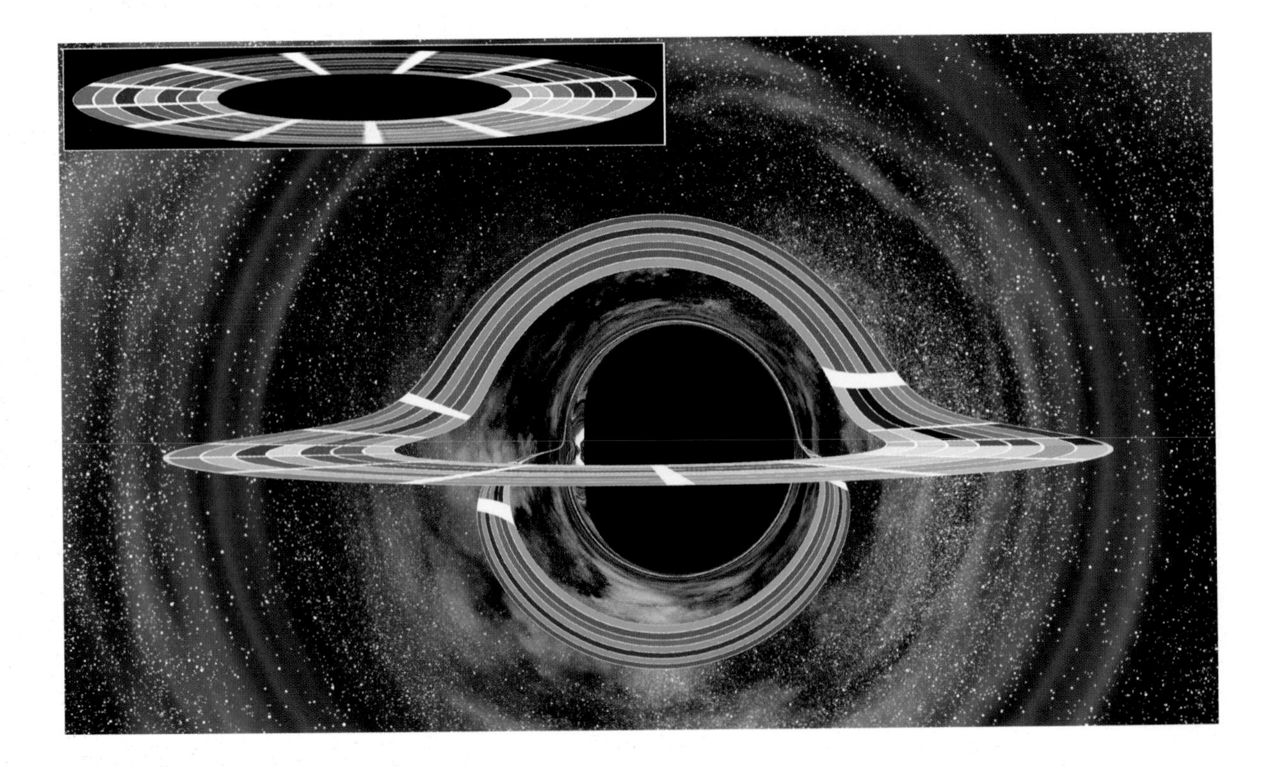

Fig. 9.7. An infinitesimally thin disk in Gargantua's equatorial plane, gravitationally lensed by Gargantua's warped space and time. Here Gargantua spins very fast. *Inset*: The disk in the absence of the black hole. *[From Eugénie von Tunzelmann's artistic team at Double Negative.]*

Inside these primary images, we see thin secondary images of the disk, wrapping over and under the shadow, near the shadow's edge. And if the picture were made much larger, you would see tertiary and higher-order images, closer and closer to the shadow.

Can you figure out why the lensed disk has the form you see? Why is the primary image wrapping under the shadow attached to the thin secondary image wrapping over it? Why are the paint swatches on the over-wrapping and under-wrapping images widened so greatly, and those on the sides squeezed? . . .

Fig. 9.8. Light rays (*red*) bring to the camera images of the back part of the accretion disk, behind Gargantua: one image above the hole's shadow, the other below the hole's shadow.

Gargantua's space whirl (space moving toward us on the left and away on the right) distorts the disk images. It pushes the disk away from the shadow on the left and toward the shadow on the right, so the disk looks a bit lopsided. (Can you explain why?)

To get further insight, Eugénie von Tunzelmann and her team replaced their variant of the color-swatch disk (Figure 9.7) with a more realistic thin accretion disk: Figure 9.9. This was much more beautiful, but it raised problems. Chris did not want his mass audience to be confused by the lopsidedness of the disk and black-hole shadow, and the shadow's flat left edge, and the complicated star-field

Fig. 9.9 Gargantua with the infinitesimally thin paint-swatch disk (Fig. 9.7) replaced by a more realistic, infinitesimally thin accretion disk. *[From Eugénie von Tunzelmann's artistic team at Double Negative.]*

patterns near that edge (discussed in Chapter 8). So he and Paul slowed Gargantua's spin to 0.6 of the maximum, making these weirdnesses more modest. (Eugénie had already omitted the Doppler shift caused by the disk's motion toward us on the left and away on the right. It would have made the disk far more lopsided: bright blue on the left and dim red on the right—totally confusing to a mass audience!)

The artistic team at Double Negative then gave the disk the texture and surface relief that we expect a real, anemic accretion disk to have, puffing it up a bit in a manner that varied from place to place. They made the disk hotter (brighter) near Gargantua and cooler (dimmer) at larger distances. They made it thicker at larger distances because it is Gargantua's tidal gravity that squeezes the disk into the equatorial plane, and tidal gravity is much weaker farther from the black hole. They added the background galaxy: many layers of artwork (dust, nebulae, stars). And they added lens flare—the haze and glare and streaks of light that would arise from scattering of the disk's bright light in a

Fig. 9.10. Gargantua and its accretion disk, with Miller's planet above the disk's left edge. The disk is so bright that the stars and nebulae are barely visible. *[From* Interstellar*, used courtesy of Warner Bros. Entertainment Inc.]*

camera lens. The results were the wonderful and compelling images in the movie (Figures 9.10 and 9.11).

Eugénie and her team also, of course, made the disk's gas orbit Gargantua, as it must to avoid falling in. When combined with gravitational lensing, the gas's orbital motion produced the impressive streaming effects in the movie—streaming effects that are hinted at by the gas's streamlines in Figure 9.11.

What a joy it was when I first saw these images! For the first time ever, in a Hollywood movie, a black hole and its disk depicted as we humans will really see them when we've mastered interstellar travel. And for the first time for me as a physicist, a realistic disk, gravitationally lensed, so it wraps over the top and bottom of the hole instead of being hidden behind the hole's shadow.

With Gargantua's disk anemic, though gorgeously beautiful, and with no jet, is Gargantua's environment truly benign? Amelia Brand thinks so . . .

Fig. 9.11 A segment of Gargantua's disk seen up close, with the *Endurance* passing over it. The black region is Gargantua, framed by the disk and with some white scattered light in the foreground. *[From* Interstellar*, used courtesy of Warner Bros. Entertainment Inc.]*

10

Accident Is the First Building Block of Evolution

Ⓣ

In *Interstellar,* upon finding Miller's planet sterile, Amelia Brand argues for going next to a planet very far from Gargantua, Edmunds' planet, instead of the closer Mann's planet: "Accident is the first building block of evolution," she tells Cooper. "But when you're orbiting a black hole, not enough can happen—it sucks in asteroids and comets, other events that would otherwise reach you. We need to go further afield."

This is one of the few spots in *Interstellar* where the characters get the science wrong. Christopher Nolan knew that Brand's argument was wrong, but he chose to retain these lines from Jonah's draft of the screenplay. No scientist has perfect judgment.

Although Gargantua tries to suck asteroids and comets into itself, and planets and stars and small black holes too, it rarely succeeds. Why?

When far from Gargantua, any object has a large angular momentum,[1] unless its orbit is headed almost directly toward the black hole.

1 The angular momentum is the object's circumferential speed multiplied by its distance from Gargantua; and this angular momentum is important because it is constant along the object's orbit, even if the orbit is complicated.

That large angular momentum produces centrifugal forces that easily overwhelm Gargantua's gravitational pull whenever the object's orbit carries it near the black hole.

A typical orbit has a form like that in Figure 10.1. The object travels inward, pulled by Gargantua's strong gravity. But before it reaches the horizon, centrifugal forces grow strong enough to fling the object back outward. This happens over and over again, almost endlessly.

The only thing that can intervene is an accidental near encounter with some other massive body (a small black hole or star or planet). The object swings around the other body on a slingshot trajectory (Chapter 7), and thereby is thrown into a new orbit around Gargantua with a changed angular momentum. The new orbit almost always has a large angular momentum, like the old one did, with centrifugal forces that save the object from Gargantua. Very rarely the new orbit carries the object almost directly toward Gargantua, with a small enough angular momentum that centrifugal forces can't win, so the object plunges through Gargantua's horizon.

Astrophysicists have carried out simulations of the simultaneous orbital motions of millions of stars around a gigantic black hole like Gargantua. Slingshots gradually change all the orbits and thereby change the density of stars (how many stars there are in some chosen volume). The star density near Gargantua does not go down; it grows.

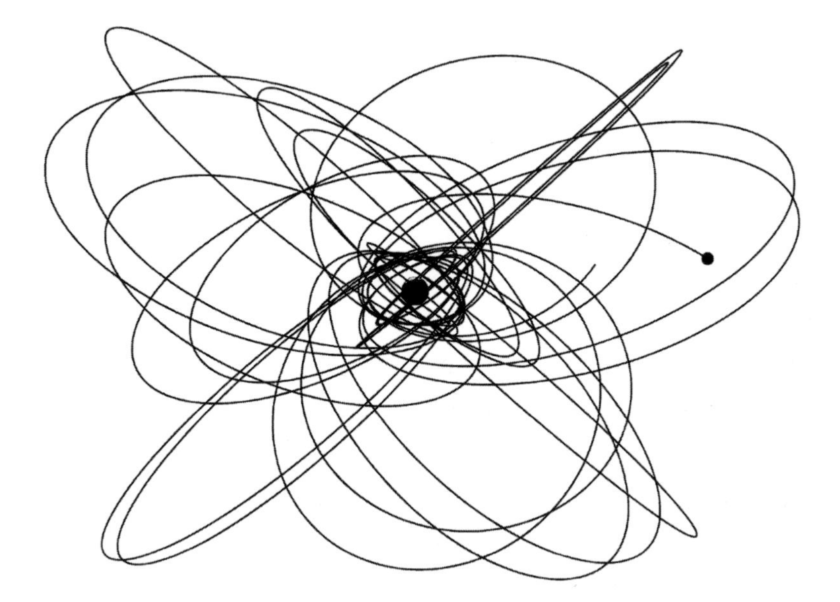

Fig. 10.1. Typical orbit of an object around a fast-spinning black hole like Gargantua. *[From a simulation by Steve Drasco.]*

And the density of asteroids and comets will also grow. Random bombardment by asteroids and comets will become more frequent, not less frequent. The environment near Gargantua will become more dangerous for individual life forms, including humans, promoting faster evolution if enough individuals survive.

With Gargantua and its dangerous environment under our belts, let's make a brief change of direction: to Earth and our solar system; to disaster on Earth and the extreme challenge of escaping disaster via interstellar travel.

III

DISASTER ON EARTH

11

Blight

In 2007, when Jonathan (Jonah) Nolan joined *Interstellar* as screenwriter, he set the movie in an era when human civilization is a pale remnant of today's and is being dealt a final blow by blight. Later, when Jonah's brother Christopher Nolan took over as director, he embraced this idea.

But Lynda Obst, Jonah, and I worried a bit about the scientific plausibility of Cooper's world, as envisioned by Jonah: How could human civilization decline so far, yet seem so normal in many respects? And is it scientifically possible that a blight could wipe out all edible crops?

I don't know much about blight, so we turned to experts for advice. I organized a dinner at the Caltech faculty club, the Athenaeum, on July 8, 2008. Great food. Superb wine. Jonah, Lynda, me, and four Caltech biologists with the right mixture of expertise: Elliot Meyerowitz, an expert on plants; Jared Leadbetter, an expert on the diverse microbes that degrade plants; Mel Simon, an expert on the cells that make up plants and how they are affected by microbes; and David Baltimore, a Nobel laureate with a broad perspective on all of biology. (Caltech is a wonderful place. Named the top university in the world by the *Times* of London in each of the last three years, it is small enough—just 300 professors, 1000 undergrads, and 1200 graduate students—that I know Caltech experts in all branches of science. It was easy to find and recruit the experts we needed for our Blight Dinner.)

As dinner began I placed a microphone at the center of our round table and recorded our two-and-a-half-hour, free-wheeling conversation. This chapter is based on that recording, but I've paraphrased what people said—and they checked and approved my paraphrasing.

Our final consensus, easily reached, is that Cooper's world is scientifically possible, *but not very likely*. It is very unlikely to happen, but it could. That's why I labeled this chapter Ⓢ for speculative.

Cooper's World

Over wine and hors d'oevres, Jonah described his vision for Cooper's world (Figure 11.1): Some combination of catastrophes has reduced the population of North America tenfold or more, and similarly on all other continents. We have become a largely agrarian society, struggling to feed and shelter ourselves. But ours is not a dystopia. Life is still tolerable and in some ways pleasant, with little amenities such as baseball continuing. However, we no longer think big. We no longer aspire to great things. We aspire to little more than just keeping life going.

Most of us think the catastrophes are finished, that we humans are securing ourselves in this new world and things may start improving. But in reality the blight is so lethal, and leaps so quickly from crop to crop, that the human race is doomed within the lifetime of Cooper's grandchildren.

What Catastrophes?

What kind of catastrophes could have produced Cooper's world? Our biologist experts offered a number of possible, but improbable, answers. Here are several:

Leadbetter: Today (2008) most people aren't growing their own food. We're dependent on a global system for growing and distributing

Fig. 11.1. Aspects of life in Cooper's world. *Top*: A baseball game with a dust storm on the horizon. *Bottom*: Cooper's home and truck after the storm. *[From* Interstellar, *used courtesy of Warner Bros. Entertainment Inc.]*

food, and for distributing water. You could imagine that system breaking down due to some biological or geophysical catastrophe. As an example on a small scale, if there was no snow in the Sierra Nevada Mountains for a few consecutive years, there would be little drinking water in Los Angeles. Ten million people would be forced to migrate, and agricultural output in California would plummet. You can easily imagine much larger scale catastrophes. In Cooper's world, with a vastly reduced population and a return to agrarian society, the production and distribution problems are lessened.

Simon: Another possible catastrophe: Over human history there has been a continual battle between us and *pathogens* (microbes that attack the human body or attack plants or other animals). We humans have developed a sophisticated immune system to deal with the pathogens that attack us directly. But the pathogens keep evolving and we're always half a step behind them. At some point there could be a catastrophe where the pathogens change so fast that our immune systems can't keep up.

Baltimore: For example, the AIDS virus could quickly evolve into a far more contagious form, one transmitted by coughing or breathing instead of sex.

Simon: The Earth's ice caps, melting due to global warming, could release a long-dormant lethal pathogen from before the last ice age.

Leadbetter: Yet another scenario: People could panic about global warming. The warming is largely caused by increasing carbon dioxide in the atmosphere. To save us, they might fertilize the Earth's oceans to produce algae that will eat much of the atmosphere's carbon dioxide via photosynthesis. A lot of iron, thrown into the oceans, could do the job. But there might be catastrophic unintended side effects. You might get some new kinds of algae that produce *toxins* (poison chemicals, not deadly life-forms) that poison the oceans. There would be a massive kill off of fish and plant life. Human civilization depends heavily on the oceans. This could be catastrophic for humans. Is it impossible? Not at all. Experiments have been done where iron was thrown locally into the ocean to produce algae—so much algae that it could be seen from space as green spots (Figure 11.2). Some of the algae that bloomed were of types never before known to science! We were lucky: the new algae were not noxious, but they might have been.

Meyerowitz: Ultraviolet light, streaming through our atmosphere's ozone hole, could mutate your enormous bloom of algae so it creates

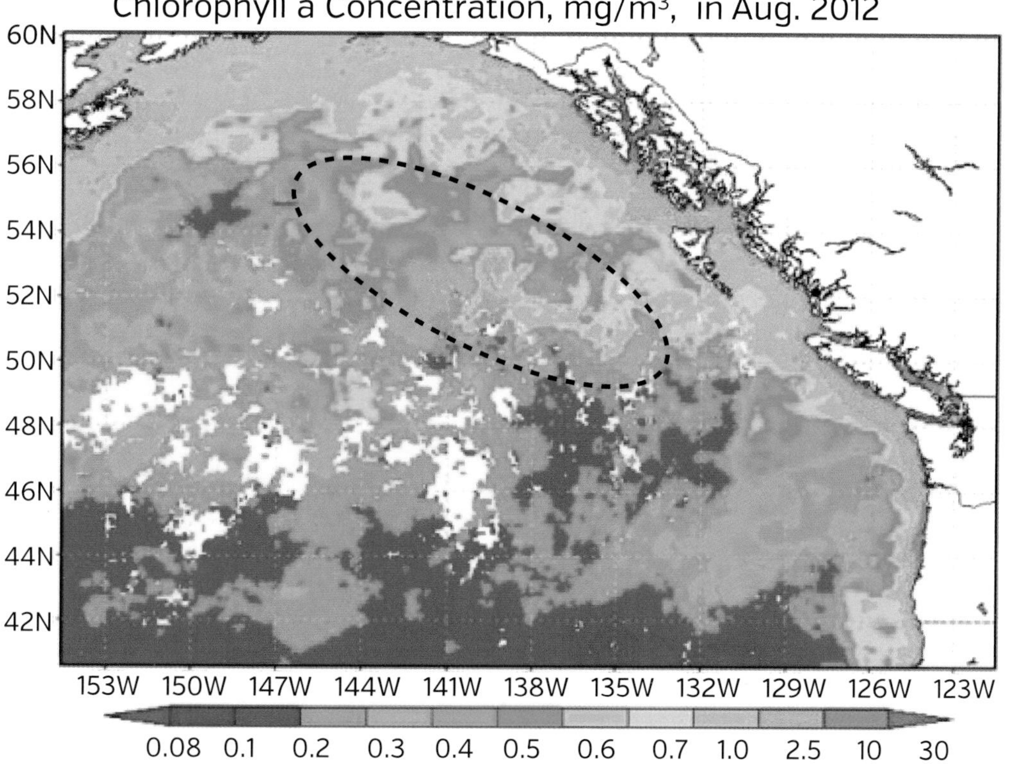

Fig. 11.2. Map of chlorophyll concentration (algae) after dumping 100 tons of iron sulphate into the ocean off the coast of British Columbia. Iron-stimulated algae growth produced the high algae concentration inside the dashed ellipse. *[From Giovanni/Goddard Earth Sciences Data and Information Services Center/NASA.]*

new pathogens. These pathogens could wipe out plants in the ocean, and then jump to land and start wiping out crops.

Baltimore: When faced with catastrophes like these, our only hope for dealing with them is advanced science and technology. If, politically, we don't invest in science and technology, or we hobble them by anti-intellectual ideologies such as denial of evolution, the very source of these catastrophes, we may find ourselves without the solutions we need.

And then there is *blight*—the consequence of many of these scenarios.

Blight

Blight is a general term for most any disease in a plant that is caused by a pathogen.

Baltimore: If you want something to wipe out humanity, there might be no better way than a blight that attacks plants. We are dependent on plants to eat. Yes, we can eat animals or fish instead, but they ate plants.

Meyerowitz: It might be sufficient for the blight just to kill off the grasses and nothing else. Grasses are the basis of most of our agriculture: rice, corn, barley, sorghum, wheat. And most animals that we eat feed on grasses.

Meyerowitz: We already live in a world where 50 percent of the food grown is destroyed by pathogens, and it's much higher than that in Africa. Fungi, bacteria, viruses, . . . they all can be pathogens. The East Coast used to be covered with chestnut trees. They are no more. They were killed by a blight. The species of banana preferred by most people in the eighteenth century was wiped out by a blight. The replacement species, the Cavendish banana, today is being threatened by blight.

Kip: I thought that blights are *specialists* that attack only one narrow group of plants and don't jump to others.

Leadbetter: There are also *generalist* blights. There seems to be a tradeoff between being a generalist that attacks many species and a specialist that attacks only a few. For the specialist blight, the lethality can be turned up really high; it can knock out, say, 99 percent of a very specific group of plants. For the generalist, the range of plants attacked is much broader, but its lethality for any one plant in that range might be much smaller. That's a pattern we see again and again in Nature.

Lynda: Could you have a generalist blight that becomes much more lethal?

Meyerowitz: Something like that has happened before. Early in the Earth's history, when cyanobacteria started making oxygen, thereby changing radically the composition of the Earth's atmosphere, they managed to kill most everything else on Earth.

Leadbetter: But the oxygen was a lethal byproduct, a poison, produced by the cyanobacteria; not a generalist pathogen.

Baltimore: We may not have seen it, but I can imagine a very lethal specialist pathogen becoming a lethal generalist. It could spread the range of plants it attacks with the help of an insect that carries it to many species. A Japanese beetle, for example, which eats something like two hundred different plant species, could infect many species with the pathogen it carries, and the pathogen might adapt to attack those species, lethally.

Meyerowitz: I can conceive of a totally lethal generalist: a pathogen that attacks chloroplasts. Chloroplasts are something that all plants have in common. They are crucial to photosynthesis (the process where a plant combines sunlight with carbon dioxide from the air, and water from its roots, to produce carbohydrates that it needs for growth). Without chloroplasts, a plant will die. Now suppose that some new pathogen evolves, for example in the oceans, that attacks chloroplasts. It could wipe out all algae and plant life in the oceans, and jump to the land where it wipes out all land plants. So everything becomes a desert. This is possible; I see nothing to prevent it. But it's not very plausible. It is unlikely ever to happen, but it could be a basis for Cooper's world.

These speculations give us a sense of the kinds of nightmare scenarios that could keep a biologist awake at night. In *Interstellar,* the focus is a lethal generalist blight running rampant over the Earth. But Professor Brand has a secondary worry: humankind's running out of oxygen to breathe.

12

Gasping for Oxygen

Early in *Interstellar* Professor Brand says to Cooper: "Earth's atmosphere is 80 percent nitrogen. We don't even breathe nitrogen. Blight does. And as it thrives, our air gets less and less oxygen. The last people to starve will be the first to suffocate. And your daughter's generation will be the last to survive on Earth."

Is there any basis in science for the Professor's prediction?

This question lies at the interface of two branches of science: biology and geophysics. So I asked the biologists at our Blight Dinner, particularly Elliot Meyerowitz, and I asked two geophysicists, Caltech professors Gerald Wasserburg (an expert on the origin and history of the Earth, Moon, and solar system) and Yuk Yung (an expert on the physics and chemistry of our Earth's atmosphere, and the atmospheres of other planets). From them, and from technical articles they pointed me to, I learned the following.

Creating and Destroying Breathable Oxygen

The oxygen we breathe is O_2: a molecule made of two oxygen atoms, bound together by electrons. There is lots of oxygen on Earth in other forms: carbon dioxide, water, minerals in the Earth's crust, to name a few. But our bodies can't use that oxygen until some organism liberates it and converts it to O_2.

The atmosphere's O_2 is *destroyed* by breathing, burning, and decay. When we breathe in O_2 our bodies combine it with carbon to form carbon dioxide, CO_2, releasing lots of energy that our bodies use. When wood is burned, the flames rapidly combine the atmosphere's O_2 with the wood's carbon to form CO_2, which generates the heat that keeps the burning going. When dead plants decay on the forest floor, their carbon is slowly combined with the atmosphere's O_2 to form CO_2 and heat.

The atmosphere's O_2 is *created* primarily by photosynthesis: chloroplasts in plants[1] (Chapter 11) use energy from sunlight to split CO_2 into C and O_2. The O_2 is liberated into the Earth's atmosphere, while the plants combine the carbon with hydrogen and oxygen from water to form the carbohydrates that they need for growth.

O_2 Destruction and CO_2 Poisoning

Suppose evolution creates a pathogen that destroys chloroplasts, as speculated by Elliot Meyerowitz at the end of the last chapter. Photosynthesis ends, not all at once, but gradually as plants die out. O_2 is no longer being created, but it is still being destroyed by breathing, burning, and decay—primarily decay, it turns out. Fortunately for the remaining humans, there is not enough decaying plant life on the Earth's surface to swallow up all the O_2.

Most of the decay will be finished after thirty years, and only about 1 percent of the O_2 will be used up. There is still plenty for Cooper's children and grandchildren to breathe, if they can find anything to eat.

But that 1 percent of the atmospheric O_2 will have been converted into carbon dioxide, which means 0.2 percent of the atmosphere will then be CO_2 (since most of the atmosphere is nitrogen). That's enough

1 Chloroplasts and photosynthesis also occur in algae, and in cyanobacteria in the ocean, both of which I treat as plant life in my simplified description. (In some sense, cyanobacteria are a form of chloroplast.)

CO_2 to make breathing unpleasant for highly sensitive people and perhaps drive the Earth's temperature up (via the greenhouse effect) by 10 degrees Celsius (18 degrees Fahrenheit)—unpleasant for everyone, to put it mildly!

To make everyone's breathing uncomfortable and induce drowsiness, ten times more atmospheric O_2 would have to be converted into CO_2; and to kill most everyone by CO_2 poisoning, an additional five times more would have to be converted, a factor of fifty in all. I have not found a plausible mechanism for this.

So is Professor Brand wrong? (Even theoretical physicists can make mistakes. Especially theoretical physicists. I know; I am one.) Probably yes, he is wrong, but conceivably no. The Professor *could* be right, but it would require geophysicists' understanding of ocean bottoms to be severely flawed.

There is undecayed organic material on the ocean bottoms as well as on land. Geophysicists estimate that the amount on ocean bottoms is about one-twentieth that on land. *If* they are wrong and there is fifty times more on the ocean bottoms than on land, and *if* there is a mechanism to quickly dredge it up, then its decay to produce CO_2 could leave everyone gasping for oxygen and dying of CO_2 poisoning.

Now, once every many thousand years, an instability triggers the ocean to turn over. Water from the surface sinks to the bottom and drives bottom water to the surface. It is conceivable that in Cooper's era there is such an overturn so vigorous that the upwelling bottom water brings with itself most of the ocean bottoms' organic material. Suddenly exposed to the atmosphere, this material could decay, converting atmospheric O_2 into lethal amounts of CO_2.

Conceivable, yes. But highly improbable on two counts: highly unlikely that there is 1000 times more undecayed ocean-bottom organic material than geophysicists think, and highly unlikely that a sufficiently vigorous oceanic overturn will occur.[2]

Nevertheless, in *Interstellar* the Earth is surely dying and humanity must find a new home. The solar system, aside from Earth, is inhospitable, so the search is on, beyond our solar system.

2 For some quantitative details and explanations of the huge uncertainties in the geophysical estimates, see *Some Technical Notes* at the end of the book.

13

Interstellar Travel

Professor Brand tells Cooper, in their first meeting, that the Lazarus missions have been sent out to search for new homes for humanity. Cooper responds, "There's no planet in our solar system that can support life, and it'd take a thousand years to reach the nearest star. That doesn't even qualify as futile. Where did you send them, Professor?"

The worse-than-futile challenge, if you don't have a wormhole, is obvious when you realize just how far it is to the nearest stars (Figure 13.1).

Distances to Nearest Stars

Ⓣ

The nearest star (other than our Sun) thought to have a habitable planet is Tau Ceti, 11.9 light-years from Earth, so traveling at light speed you would need 11.9 years to reach it. If there are any habitable planets closer than that, they can't be much closer.

To get some sense of just how far Tau Ceti *is* compared to more familiar things, let's scale its distance down enormously. Imagine it

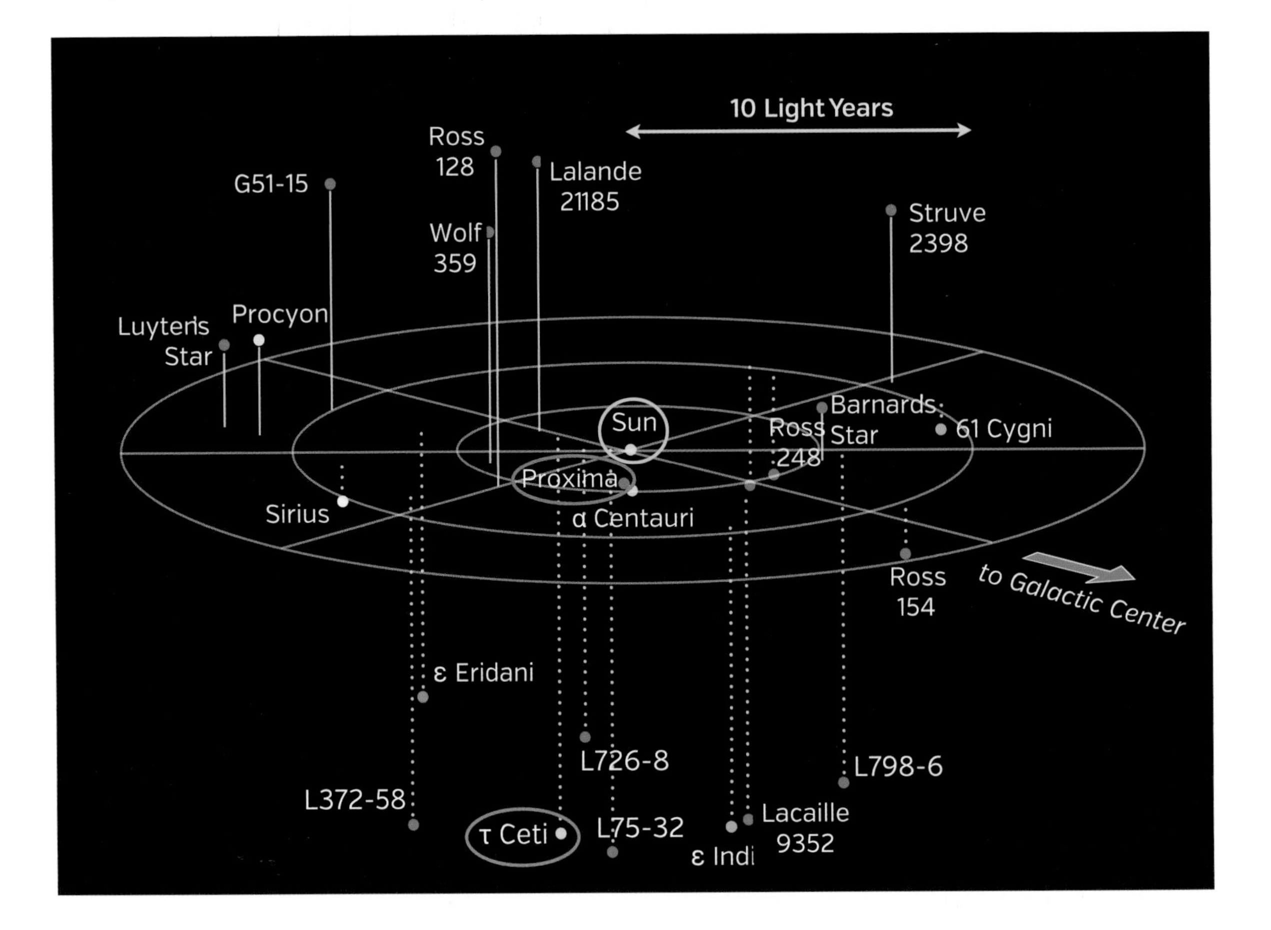

Fig. 13.1. All the stars within 12 light-years of Earth. The Sun, Proxima Centauri, and Tau Ceti are circled in yellow, purple, and red. *[I adapted this map from Richard Powell's www.atlasoftheuniverse.com.]*

as the distance from New York City to Perth, Australia, about halfway around the world.

The very nearest star other than the Sun is Proxima Centauri, 4.24 light-years from Earth, but there is no evidence it has habitable planets. With Tau Ceti's distance imagined as New York to Perth, then Proxima Centauri's is like New York to Berlin. It's not a lot closer than Tau Ceti!

For comparison, the most distant unmanned spacecraft that humans have sent into interstellar space is *Voyager 1*, now about 18 light-hours from Earth. It has been traveling for thirty-seven years to get there. With Tau Ceti's distance imagined as New York to Perth, then Earth to *Voyager 1* is about 3 kilometers (2 miles): the distance from the Empire State Building to the southern end of Greenwich Village. That's hugely less than New York to Perth.

The Earth to Saturn is even smaller: 200 meters, two east-west blocks in New York City, from the Empire State Building to Park Avenue. The Earth to Mars is just 20 meters; and the Earth to the Moon (the greatest distance humans have ever yet traveled) is just 7 centimeters—about two and a half inches!

Compare what we have achieved in going to the Moon, *two and a half inches,* with the challenge of going *halfway around the world.* That's the leap of technology required to take humans to habitable planets outside our solar system!

Travel Times with Twenty-First-Century Technology

Ⓣ

Voyager 1 is traveling out of the solar system at 17 kilometers per second, having been boosted by gravitational slingshots around Jupiter and Saturn. In *Interstellar,* the *Endurance* travels from Earth to Saturn in two years, at an average speed of about 20 kilometers per second. The fastest speed I think rocket technology plus solar system slingshots are likely to achieve in this, the twenty-first century, is about 300 kilometers per second.

At that 300 kilometers per second, we would need 5000 years to reach Proxima Centauri and 13,000 years to reach Tau Ceti. Not a pleasant prospect!

To get there far faster in the tweny-first century, you need something like a wormhole (Chapter 14).

Far-Future Technology

(EG)

Technically savvy scientists and engineers have put much effort into conceiving far-future technologies that might make possible near-light-speed travel. You can learn a lot about their ideas by browsing the web. It will take many centuries for humans to make any of those ideas real, I think. But they do convince me that ultra-advanced civilizations are likely to travel between the stars at a tenth the speed of light or faster.

Here are three far-out examples of near-light-speed propulsion that intrigue me.

Thermonuclear Fusion

EG

Thermonuclear fusion is the most conventional of the three ideas. R&D to develop controlled-fusion power plants on Earth was initiated in the 1950s, and full success will not come until the 2050s. A full century of R&D! That's a realistic measure of the difficulties.

And what will fusion power plants in 2050 mean for spacecraft propulsion by fusion? The most practical designs may achieve 100 kilometers per second, and conceivably 300 kilometers per second by the end of this century. A whole new approach to harnessing fusion will be required for reaching near light speed.

A simple calculation shows fusion's possibility: When two deuterium (heavy hydrogen) atoms are fused to form a helium atom, 0.0064 (nearly 1 percent) of their rest mass gets converted into energy. If this were all transformed to kinetic energy (energy of motion) of the helium atom, the atom would move at about one-tenth the speed of light.[1] This suggests that, if we could convert all the fusion energy of deuterium fuel into ordered motion of a spacecraft, we could achieve a spacecraft speed of roughly 1/10 the speed of light—and somewhat higher if we are clever.

In 1968 Freeman Dyson, a brilliant physicist for whom I have great respect, described and analyzed a crude propulsion system that, in the hands of a sufficiently advanced civilization, could achieve this.

1 The kinetic energy is $Mv^2/2$, where M is the helium atom's mass and v is its speed. Equate this to the energy released, $0.0064\,Mc^2$, where c is the speed of light. (I used Einstein's famous result that when you convert mass into energy, the energy you get out is the mass multiplied by the square of the speed of light.) The result from equating these two formulas is $v^2 = 2 \times 0.0064c^2$, which means v is close to $c/10$.

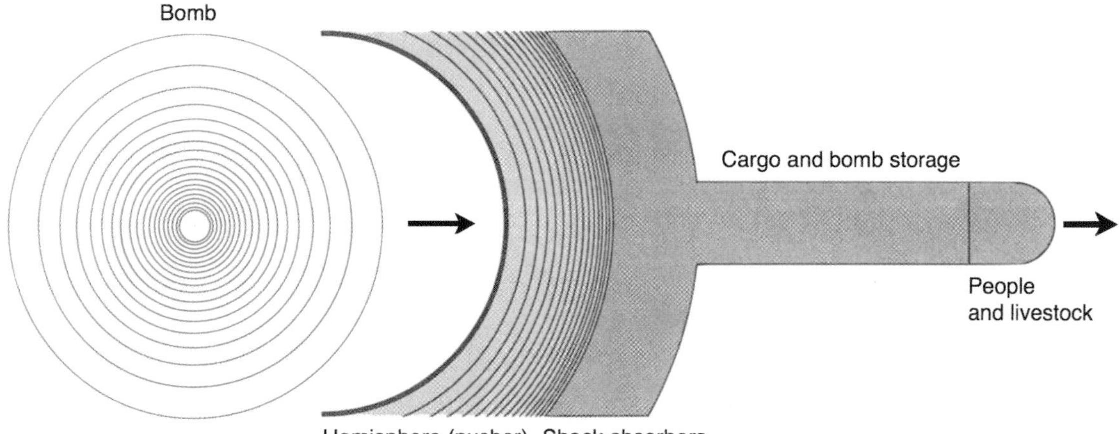

Fig. 13.2. Freeman Dyson's bomb-powered propulsion system. *[From Dyson (1968).]*

Thermonuclear bombs ("hydrogen bombs") are detonated just behind a hemispherical shock absorber that is 20 kilometers in diameter (Figure 13.2). The bomb debris pushes the ship forward, achieving, in Dyson's most optimistic estimate, a speed one-thirtieth that of light. A less crude design could do somewhat better. In 1968 Dyson estimated that such a propulsion system would not be practical any sooner than the late twenty-second century, 150 years from now. I think that's overly optimistic.

Laser Beam and Light Sail

Ⓢ

In 1962 Robert Forward, another physicist whom I respect, wrote a short article in a popular magazine about a spacecraft with a sail, pushed by a distant, focused laser beam (Forward 1962). In a 1984 technical article, he made this concept more sophisticated and precise (Figure 13.3.)

An array of solar-powered lasers in space or on the Moon generates a laser beam with 7.2 terawatts of power (about twice the total power consumption of the United States in 2014!). This beam is focused, by a Fresnel lens 1000 kilometers in diameter. It is focused onto a distant sail, 100 kilometers in diameter and weighing about 1000 metric tons, that is attached to a less massive spacecraft. (The beam direction must be accurate to about a millionth of an arcsecond.) The beam's

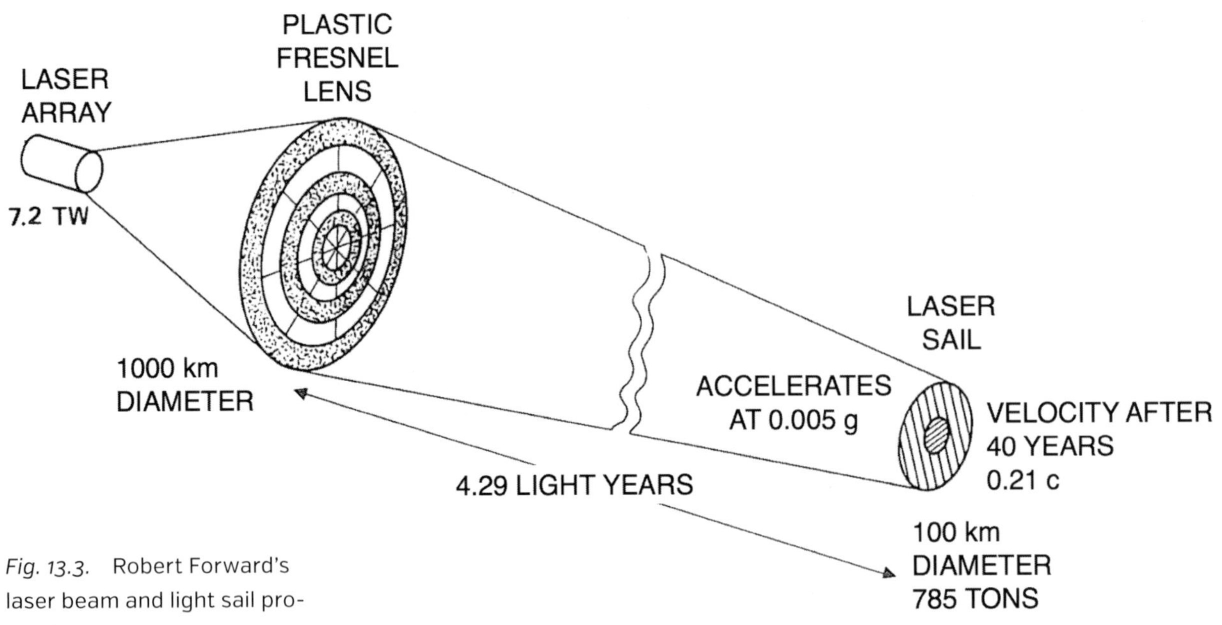

Fig. 13.3. Robert Forward's laser beam and light sail propulsion system. *[From Forward (1984).]*

light pressure pushes the sail and spacecraft up to about a fifth the speed of light halfway through a forty-year trip to Proxima Centauri. A modification of this scheme then slows the ship down during the second half of the trip, so it arrives at its destination with a speed low enough to rendezvous with a planet. (Can you figure out how the slow down is achieved?)

Forward, like Dyson, imagined his scheme practical in the twenty-second century. When I look at the technical challenges, I think longer.

Gravitational Slingshots in a Black-Hole Binary

My third example is my own wild—very wild!—variant of an idea due to Dyson (1963).

Suppose you want to fly across much of the universe (not just interstellar travel, but intergalactic travel) at near light speed in a few years of your own life. You can do so with the aid of two black holes that are orbiting each other, a *black-hole binary*. They must be in a highly elliptical orbit and must be large enough that their tidal forces do not destroy your ship.

Using chemical or nuclear fuel, you navigate your ship into an orbit that comes close to one of the black holes: a so-called zoom-whirl orbit (Figure 13.4). Your ship zooms close to the hole, whirls around it a few times, and then, when the hole is traveling nearly directly toward its companion, the ship zooms out, crosses over to the companion hole, and slides into a whirl around it. If the two holes are still headed toward each other, the whirl is brief: you zoom back toward the first hole. If the holes are no longer headed toward each other, the whirl is much longer; you must park yourself in orbit around the second hole until the holes are again headed toward each other, and then launch back toward the first hole. In this way, always traveling between holes only when the holes are approaching each other, your ship gets boosted to higher and higher speeds, approaching as close as you wish to the speed of light if the binary is sufficiently elliptical.

It is a remarkable fact that you only need a small amount of rocket fuel to control how long you linger around each hole. The key is to navigate onto the hole's critical orbit, and there perform your controlled whirl. I discuss the critical orbit in Chapter 27. For now, suffice it to say that this is a highly *unstable* orbit. It is rather like riding a motorcycle around a very smooth volcano rim. If you balance delicately, you can stay on the rim as long as you want. When you wish to leave, a slight turn of the bike's front wheel will send you careening off the rim. When you want to leave the critical orbit, a slight rocket thrust will enable centrifugal forces to take over and send your ship careening toward the other black hole.

Once you are as close to the speed of light as you wish, you can

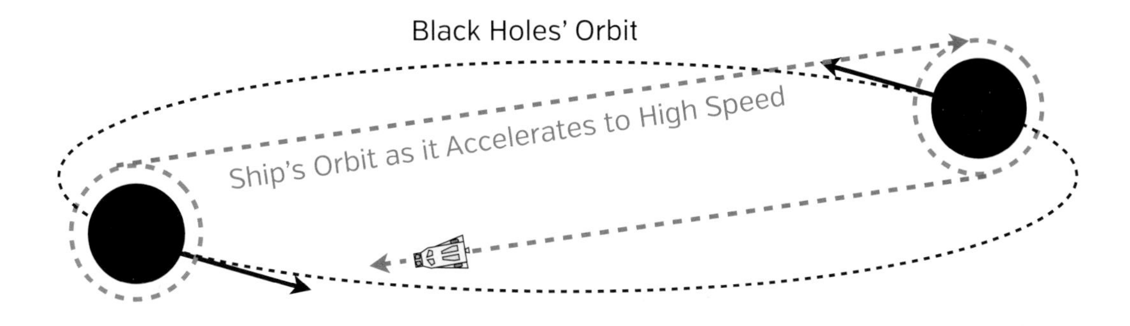

Fig. 13.4. Zoom-whirl orbit brings a spacecraft up to near light speed.

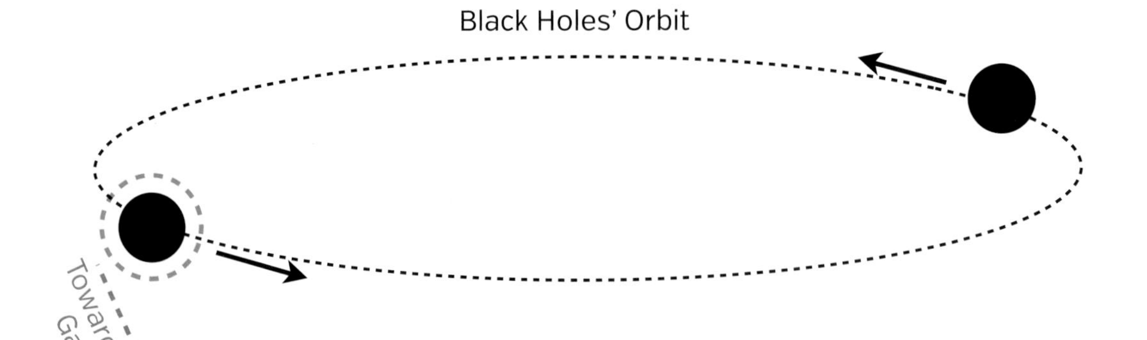

Fig. 13.5. Launching off a critical orbit toward a distant galaxy.

launch yourself off a critical orbit toward your target galaxy in the distant universe (Figure 13.5).

The trip may be long; as much as 10 billion light-years' distance. But when you move at near light speed, your time flows far more slowly than on Earth. If you are close enough to light speed, you can make it to your target in a few years or less, as measured by you—slowing down with the aid of a highly elliptical black-hole binary at your target, if you can find one! See Figure 13.6.

You can return home by the same method. But your homecoming may not be pleasant. Billions of years will have passed at home, while you have aged only a few years. Imagine what you find.

These types of slingshots could provide a means for spreading a civilization across the great reaches of intergalactic space. The principal obstacle (perhaps insurmountable!) is finding, or making, the needed black hole binaries. The launch binary might not be a problem if you are a sufficiently advanced civilization, but the slow-down binary is another matter.

What happens to you if there is no slow-down binary, or there is

Fig. 13.6. Slowing down by slingshots in a target black-hole binary.

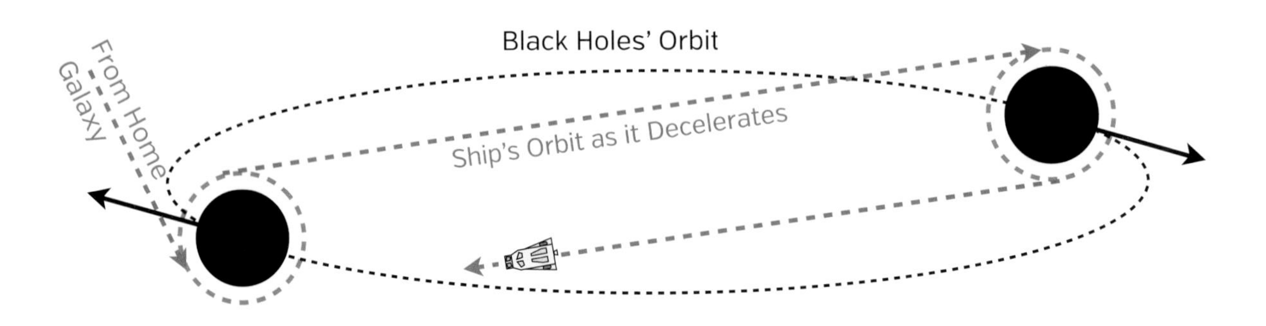

one, but your aim is bad and you miss it? This is a tricky question because of the expansion of the universe. Think about it.

As exciting as these three far-future propulsion systems may seem, they truly *are* far future. Using twenty-first-century technology, we are stuck with thousands of years to reach other solar systems. The only hope (an exceedingly faint hope) for faster interstellar travel, in the event of an earthly disaster, is a wormhole like that in *Interstellar,* or some other extreme form of spacetime warp.

IV

THE WORMHOLE

14

Wormholes

How Wormholes Got Their Name

Ⓣ

My mentor, John Wheeler, gave astrophysical wormholes their name. He based it on wormholes in apples (Figure 14.1). For an ant walking on an apple, the apple's surface is the entire universe. If the apple is threaded by a wormhole, the ant has two ways to get from the top to the bottom: around the outside (through the ant's universe) or down the wormhole. The wormhole route is shorter; it's a shortcut from one side of the ant's universe to the other.

Fig. 14.1. An ant explores a wormhole-endowed apple.

The apple's delicious interior, through which the wormhole passes, is not part of the ant's universe. It is a three-dimensional bulk or hyperspace (Chapter 4). The wormhole's wall can be thought of as part of the ant's universe. It has the same dimensionality as the universe (two dimensions) and it joins onto the universe (the apple's surface) at the wormhole's entrance. From another viewpoint, the wormhole's wall is *not* part of the ant's universe; it is just a shortcut by which the ant can travel across the bulk, from one point in its universe to another.

Flamm's Wormhole

Ⓣ

In 1916, just one year after Einstein formulated his general relativistic laws of physics, Ludwig Flamm in Vienna discovered a solution of Einstein's equations that describes a wormhole (though he did not call it that). We now know that Einstein's equations allow many kinds of wormholes (wormholes with many different shapes and behaviors), but Flamm's is the only one that is precisely spherical and contains no gravitating matter. When we take an equatorial slice through Flamm's wormhole, so it and our universe (our brane) have just two dimensions rather than three, and when we then view our universe and the wormhole from the bulk, they look like the left part of Figure 14.2.

With one of our universe's dimensions lost from the picture, you must think of yourself as a two-dimensional creature confined to move on the bent sheet or on the wormhole's two-dimensional wall. There are two routes for travel from location $\mathcal{A}$ in our universe to location $\mathcal{B}$: the short route (dashed blue curve) down the wormhole's wall, or the long route (dashed red curve) along the bent sheet, our universe.

Of course, our universe is really three dimensional. The concentric circles in the left part of Figure 14.2 are really the nested green spheres shown to the right. As you enter the wormhole along the blue path from location $\mathcal{A}$, you pass through spheres that get smaller and smaller. Then the spheres, though nested inside each other, cease changing circumference. And then, as you exit the wormhole toward location $\mathcal{B}$, the spheres get larger and larger.

For nineteen years, physicists paid little attention to Flamm's outrageous solution of Einstein's equations, his wormhole. Then in 1935 Einstein himself and fellow physicist Nathan Rosen, unaware of

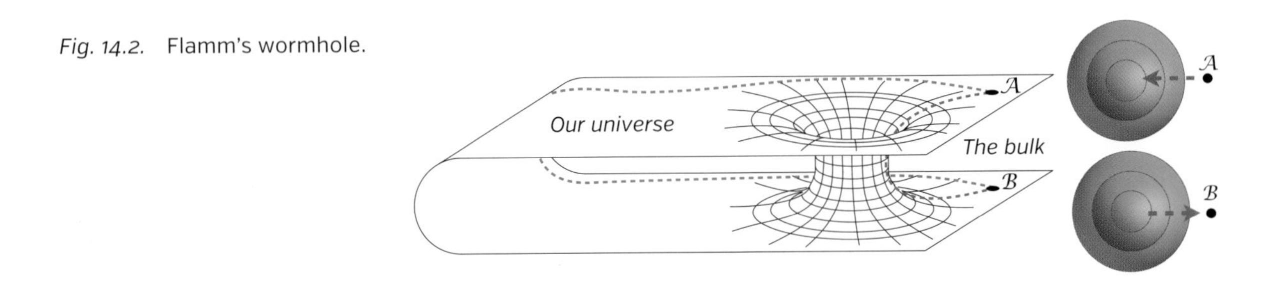

Fig. 14.2. Flamm's wormhole.

Flamm's work, rediscovered Flamm's solution, explored its properties, and speculated about its significance in the real world. Other physicists, also unaware of Flamm's work, began to call his wormhole the "Einstein-Rosen bridge."

Wormhole Collapse

Ⓣ

It is often difficult to extract, from the mathematics of Einstein's equations, a full understanding of their predictions. Flamm's wormhole is a remarkable example. From 1916 until 1962, nearly a half century, physicists thought that the wormhole is static, forever unchanging. Then John Wheeler and his student Robert Fuller discovered otherwise. Looking much more closely at the mathematics, they discovered that the wormhole is born, expands, contracts, and dies, as shown in Figure 14.3.

Initially, in picture (a), our universe has two singularities. As time passes, the singularities reach out to each other through the bulk and meet to create the wormhole (b). The wormhole expands in circumference, (c) and (d), then shrinks and pinches off (e), leaving behind the two singularities (f). The birth, expansion, shrinkage, and pinch-off happen so quickly that nothing, not even light, has time to travel

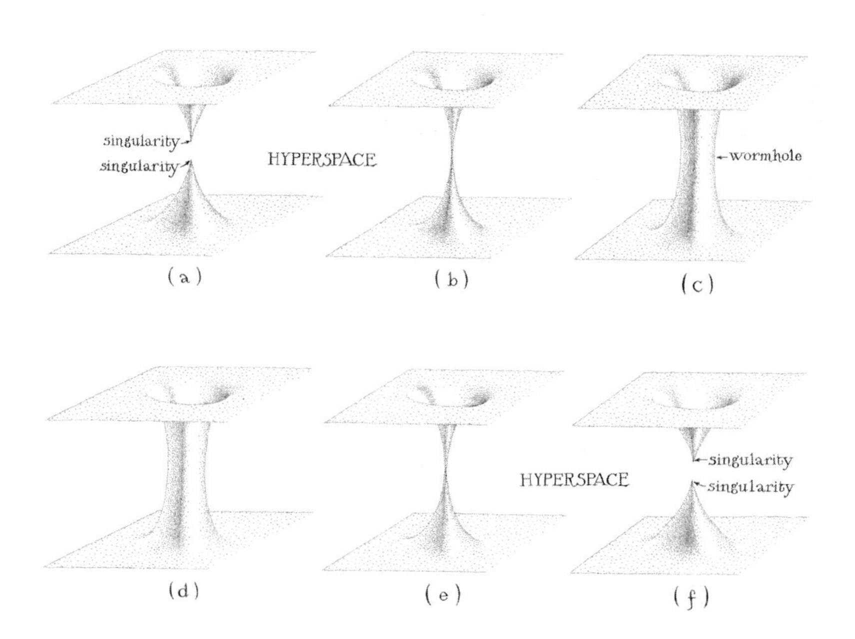

Fig. 14.3. Dynamics of Flamm's wormhole (the Einstein-Rosen bridge). *[Drawing by Matt Zimet based on a sketch by me; from my book* Black Holes & Time Warps: Einstein's Outrageous Legacy.*]*

through the wormhole from one side to the other. Anything or anyone that attempts the trip will get destroyed in the pinch-off!

This prediction is inescapable. If the universe were ever, somehow, to develop a spherical wormhole that contains no gravitating matter, this is how the wormhole would behave. Einstein's relativistic laws dictate it.

Wheeler was not dismayed by this conclusion. On the contrary, he was pleased. He regarded singularities (places where space and time are infinitely warped) as a "crisis" for the laws of physics. And crises are wonderful tutors. By probing wisely, we can get great insights into the physical laws. To this I return in Chapter 26.

Contact

Ⓣ

Fast-forward a quarter century, to May 1985: a phone call from Carl Sagan asking me to critique the relativistic science in his forthcoming novel *Contact*. I happily agreed. We were close friends, I thought it would be fun, and, besides, I still owed him one for introducing me to Lynda Obst.

Carl sent me his manuscript. I read it and I loved it. But there was one problem. He sent his heroine, Dr. Eleanor Arroway, through a black hole from our solar system to the star Vega. But I knew that a black-hole interior *cannot* be a route from here to Vega or to anywhere else in our universe. After plunging through the black hole's horizon, Dr. Arroway would get killed by its singularity. To reach Vega fast, she needed a wormhole, not a black hole. But a wormhole that does *not* pinch off. A *traversable* wormhole.

So I asked myself, What do I have to do to Flamm's wormhole to save it from pinching off; to hold it open, so it can be traversed? A simple thought experiment gave me the answer.

Suppose you have a wormhole that is spherical like Flamm's, but unlike Flamm's it does not pinch off. Send a light beam into the wormhole, radially. Since all the beam's light rays travel radially, the beam must have the shape shown in Figure 14.4. It is converging (its cross-sectional area is decreasing) as it enters the wormhole, and it is diverging (its area is increasing) as it leaves the wormhole. The wormhole has bent the light rays outward, as would a diverging lens.

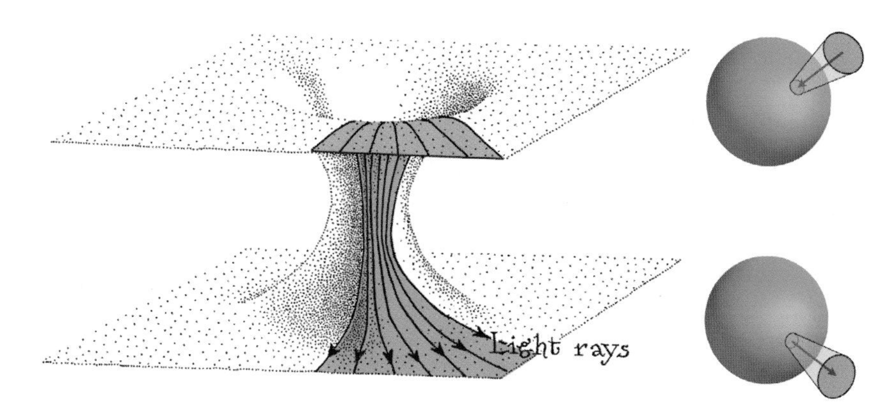

Fig. 14.4. A radial light beam traveling through a spherical, traversable wormhole. *Left*: As seen from the bulk with one space dimension removed. *Right*: As seen in our universe. *[Adapted from a drawing by Matt Zimet based on a sketch by me; from my book* Black Holes & Time Warps: Einstein's Outrageous Legacy.*]*

Now, gravitating bodies such as the Sun or a black hole bend rays inward (Figure 14.5). They can't bend rays outward. To bend light rays outward, a body must have negative mass (or equivalently, negative energy; recall Einstein's equivalence of mass and energy). From this fundamental fact, I concluded that any traversable, spherical wormhole must be threaded by some sort of material that has negative energy. At least the material's energy must be negative as seen by the light beam, or by anything or anyone else that travels through the wormhole at nearly the speed of light.[1] I call such material "exotic matter." (I later learned that, according to Einstein's relativistic laws, *any* wormhole, spherical or not, is traversable only if it is threaded by exotic matter. This follows from a theorem proved in 1975 by Dennis Gannon at the University of California at Davis. Being somewhat illiterate, I was unaware of Gannon's theorem.)

Now, it is an amazing fact that exotic matter *can* exist, thanks to weirdnesses in the laws of quantum physics. Exotic matter has even been made in physicists' laboratories, in *tiny* amounts, between two

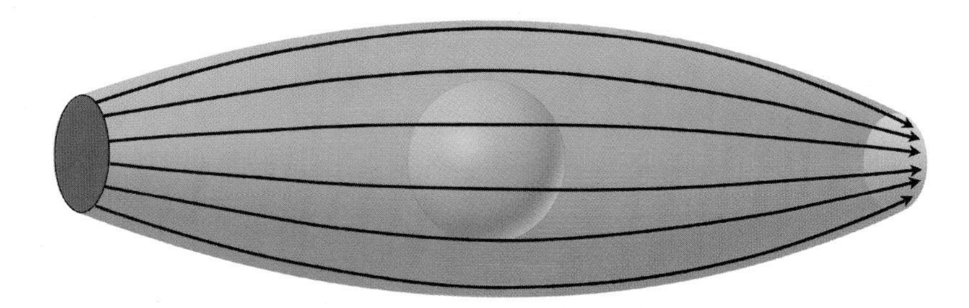

Fig. 14.5. The Sun or a black hole bends a beam of light inward.

1 Energy is weird in relativistic physics; the energy one measures depends on how fast one moves and in what direction.

closely spaced electrically conducting plates. This is called the Casimir effect. However, it was very unclear to me in 1985 whether a wormhole can contain *enough* exotic matter to hold it open. So I did two things.

First, I wrote a letter to my friend Carl suggesting that he send Eleanor Arroway to Vega through a wormhole rather than a black hole, and I enclosed a copy of the calculations by which I had shown that the wormhole must be threaded by exotic matter. Carl embraced my suggestion (and wrote about my equations in the acknowledgment of his novel). And that is how wormholes entered modern science fiction—novels, films, and television.

Second, with two of my students, Mark Morris and Ulvi Yurtsever, I published two technical articles about traversable wormholes. In our articles, we challenged our physicist colleagues to figure out whether the combined quantum laws and relativistic laws permit a very advanced civilization to collect enough exotic matter inside a wormhole to hold it open. This triggered a lot of research by a lot of physicists; but today, nearly thirty years later, the answer is still unknown. The preponderance of the evidence suggests that the answer may be NO, so traversable wormholes are impossible. But we are still far from a final answer. For details, check out *Time Travel and Warp Drives* by my physicist colleagues Allen Everett and Thomas Roman (Everett and Roman 2012).

What Does a Traversable Wormhole Look Like?

(EG)

What does a traversable wormhole look like to people like us who live in our universe? I can't answer definitively. If a wormhole *can* be held open, the precise details of *how* remain a mystery, so the precise details of the wormhole's shape are unknown. For black holes, by contrast, Roy Kerr has given us the precise details, so I can make the firm predictions described in Chapter 8.

So for wormholes, I can make only an educated guess, but one in which I have considerable confidence. Hence the symbol (EG) on this section's header.

Mouth in California Desert

Mouth in Dublin

Fig. 14.6. The images seen through a wormhole's two mouths. *[Left photo by Catherine MacBride; right photo by Mark Interrante.]*

Imagine we have a wormhole here on Earth, stretching through the bulk from Grafton Street in Dublin, Ireland, to the desert in Southern California. The distance through the wormhole might be only a few meters.

The entrances to the wormhole are called "mouths." You are sitting in a sidewalk cafe alongside the Dublin mouth. I am standing in the desert beside the California mouth. Both mouths look rather like crystal balls. When I look into my California mouth, I see a distorted image of Grafton Street, Dublin (Figure 14.6). That image is brought to me by light that travels through the wormhole from Dublin to California, rather like light traveling through an optical fiber. When you look into your Dublin mouth, you see a distorted image of Joshua trees (cactus trees) in the California desert.

Can Wormholes Exist Naturally, as Astrophysical Objects?

In *Interstellar,* Cooper says, "A wormhole isn't a naturally occurring phenomenon." I agree with him completely! If traversable wormholes are allowed by the laws of physics, I think it extremely unlikely they can exist naturally, in the real universe. I must confess, though, that this is little more than a speculation, not even an educated guess.

Maybe a highly educated speculation, but speculation nonetheless, so I labeled this section △S.

Why am I so pessimistic about natural wormholes?

We see no objects in our universe that could become wormholes as they age. By contrast, astronomers see huge numbers of massive stars that will collapse to form black holes when they have exhausted their nuclear fuel.

On the other hand, there is reason to hope that wormholes *do* exist naturally on submicroscopic scales in the form of "quantum foam" (Figure 14.7). This foam is a hypothesized network of wormholes that is continually fluctuating in and out of existence in a manner governed by the ill-understood laws of quantum gravity (Chapter 26). The foam is probabilistic in the sense that, at any moment, there is a certain probability the foam has one form and also a probability that it has another form, and these probabilities are continually changing. And the foam is truly tiny: the typical length of a wormhole would be the so-called Planck length, 0.000000000000000000000000000 000001 centimeters; a hundredth of a billionth of a billionth the size of the nucleus of an atom. That's small!!

Back in the 1950s John Wheeler gave persuasive arguments for quantum foam, but there is now evidence that the laws of quantum gravity might suppress the foam and might even prevent it from arising.

If quantum foam *does* exist, I hope there is a natural process by which some of its wormholes can spontaneously grow to human size or bigger, and even did so during the extremely rapid "inflationary" expansion of the universe, when the universe was very, very young.

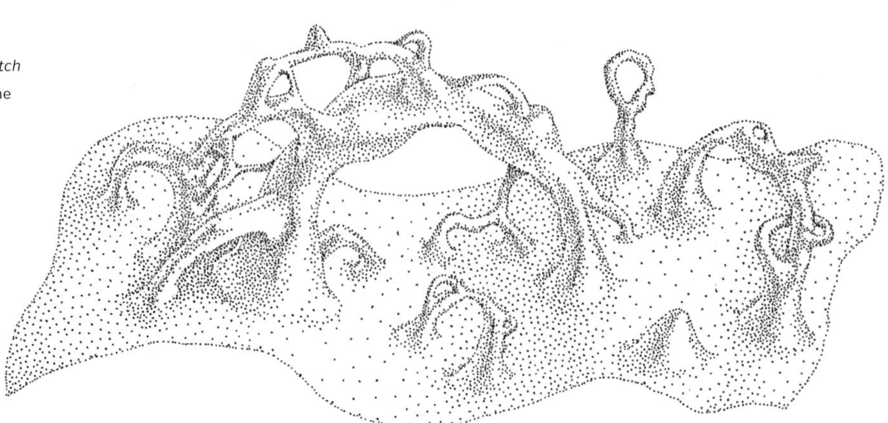

Fig. 14.7. Quantum foam. *[Drawing by Matt Zimet based on a sketch by me; from my book* Black Holes & Time Warps: Einstein's Outrageous Legacy.*]*

However, we physicists have no hint of any evidence at all that such natural enlargement can or did occur.

The other tiny hope for natural wormholes is the big bang creation of the universe. It is conceivable, but very unlikely, that traversable wormholes could have formed in the big bang itself. Conceivable for the simple reason that we don't understand the big bang well at all. Unlikely because nothing we do know about the big bang gives any hint that traversable wormholes might form there.

Can Wormholes Be Created by an Ultra-Advanced Civilization?

Ⓢ

An ultra-advanced civilization is my only serious hope for making traversable wormholes. But it would face huge obstacles, so I'm pessimistic.

One way to make a wormhole, where previously there were none, is to extract it from the quantum foam (if the foam exists), enlarge it to human size or larger, and thread it with exotic matter to hold it open. This seems like a pretty tall order, even for an ultra-advanced civilization, but perhaps only because we don't understand the quantum gravity laws that control the foam, the extraction, and the earliest stages of enlargement (Chapter 26). Of course, we don't understand exotic matter very well either.

At first sight, making a wormhole seems easy (Figure 14.8). Just push a piece of our brane (our universe) downward in the bulk to create a thimble, fold our brane around in the bulk, tear a hole in our brane just below the thimble, tear a hole in the thimble itself, and sew the tears together. *Just!*

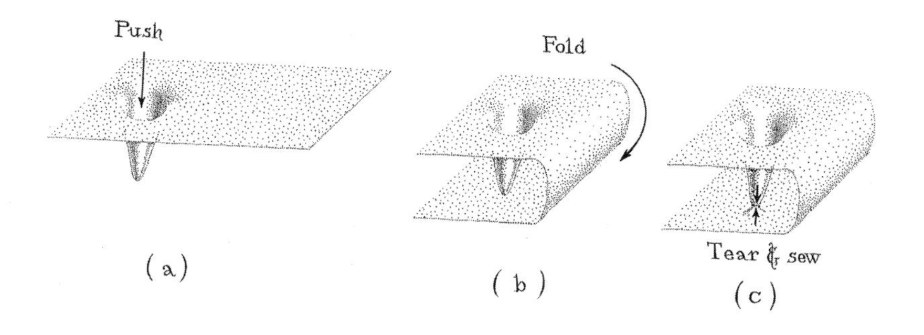

Fig. 14.8. Creating a wormhole. *[Drawing by Matt Zimet based on a sketch by me; from my book* Black Holes & Time Warps: Einstein's Outrageous Legacy.*]*

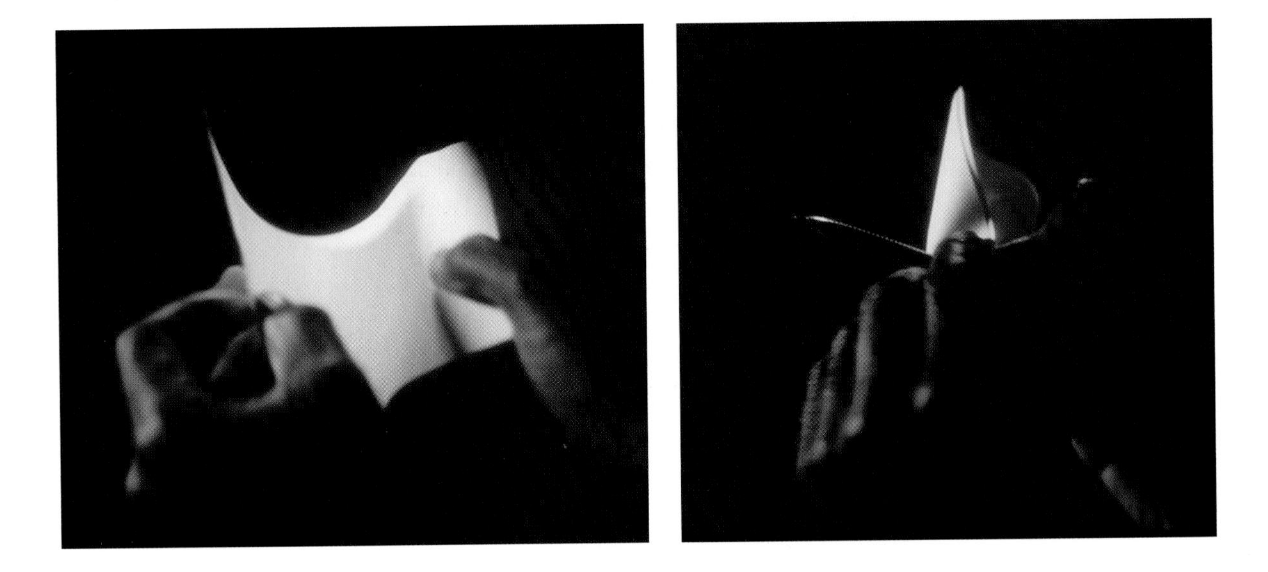

Fig. 14.9. Romilly explaining wormholes. *Left*: He bends a sheet of paper. *Right*: He punches a pencil (the wormhole) through the paper, joining its two edges. *[From* Interstellar, *used courtesy of Warner Bros. Entertainment Inc.]*

In *Interstellar,* Romilly demonstrates the same thing with a sheet of paper and a pencil (Figure 14.9). As easy as this may look from the outside, playing with pencils and paper, it is horrendously daunting when the sheet is our brane and these manipulations must be carried out from within the brane, by a civilization that lives in our brane. In fact, I have no idea how to execute any of these maneuvers from inside our brane except the first, creating a thimble in our brane (which requires only a very dense mass, such as a neutron star). Moreover, if it is possible at all to tear holes in our brane, it can only be done with the help of the laws of quantum gravity. Einstein's relativistic laws forbid tearing our brane, so the only hope is to make the tear where his laws fail, in a realm of quantum gravity. We then are back to the domain of terra almost incognita (Figure 3.2).

The Bottom Line

Ⓢ

I doubt the laws of physics permit traversable wormholes, but this may be pure prejudice. I could be wrong. If they *can* exist, I doubt very much that they can form naturally in the astrophysical universe. My only real hope for forming them is artificially, in the hands of an ultra-advanced civilization. But we are extremely ignorant of how such a civilization could do it. And it appears more than daunting, at least from inside our brane (our universe), even for the most advanced of civilizations.

In *Interstellar,* however, the wormhole is thought to have been made, held open, and placed near Saturn by a civilization that lives in the bulk, a civilization whose beings have four space dimensions, like the bulk.

This is terra *extremely* incognita. Nevertheless, I discuss bulk beings in Chapter 22. In the meantime let's talk about the wormhole in *Interstellar.*

15

Visualizing *Interstellar*'s Wormhole

The wormhole in *Interstellar* is thought to have been constructed by an ultra-advanced civilization, most likely one that lives in the bulk. In this spirit, when laying foundations for visualizing *Interstellar*'s wormhole, Oliver James[1] and I pretended we were ultra-advanced engineers. We assumed that wormholes *are* allowed by the laws of physics. We assumed the wormhole's builders had all the exotic matter they needed to hold the wormhole open. We assumed the builders could warp space and time in whatever way we wished them to, inside and around the wormhole. These are pretty extreme assumptions, so I labeled this chapter (S) for speculation.

The Wormhole's Gravity and Time Warping

Christopher Nolan wanted the wormhole to have a mild gravitational pull. Strong enough to hold the *Endurance* in orbit around itself, but

1 Recall that Oliver James, chief scientist at Double Negative, wrote the computer code that underlies *Interstellar*'s visualizations of wormholes and black holes; see Chapters 1 and 8.

weak enough that a modest rocket blast would slow the *Endurance*, letting it drop gently into the wormhole. This meant a gravitational pull much less than the Earth's.

Einstein's law of time warps tells us that the slowing of time inside the wormhole is proportional to the strength of the wormhole's gravitational pull. With that pull weaker than the Earth's pull, the slowing of time must be smaller than on Earth, which is only a part in a billion (that is, one second of slowing during a billion seconds of time, thirty years). Such slowing is so tiny that Oliver and I paid no attention to it at all when designing the wormhole.

"Handles" for Adjusting the Wormhole's Shape

The ultimate decision about the wormhole's shape was in the hands of Christopher Nolan (the director) and Paul Franklin (the visual-effects supervisor). My task was to give Oliver and his colleagues at Double Negative "handles" (or in technical language, "parameters") that they could use to adjust the shape. They then simulated the wormhole's appearance for various adjustments of the handles and showed the simulations to Chris and Paul, who chose the one that was most compelling.

I gave the wormhole's shape three handles—three ways to adjust the shape (Figure 15.1).

The first handle is the wormhole's *radius* as measured by an ultra-advanced engineer looking in from the bulk (analog of Gargantua's radius). If we multiply that radius by $2\pi = 6.28318...$, we get the wormhole's circumference as measured by Cooper when he pilots the *Endurance* around or through it. Chris chose the radius before I began to work. He wanted the wormhole's gravitational lensing of stars to be barely visible from Earth with the best large-telescope technology then available to NASA. That fixed the radius at about a kilometer.

Fig. 15.1. A wormhole viewed from the bulk and my three handles for adjusting its shape. (The inset on the left is the same wormhole, viewed from farther away in the bulk so we see its outer parts.)

The second handle is the wormhole's *length,* as measured equally well by Cooper or by an engineer in the bulk.

The third handle determines how strongly the wormhole lenses the light from objects behind it. The details of the lensing are fixed by the shape of space near the wormhole's mouths. I chose that shape similar to the shape of space outside the horizon of a nonspinning black hole. My chosen shape has just one adjustable handle: the width of the region that produces strong lensing. I call this the *lensing width*[2] and depict it in Figure 15.1.

How the Handles Influence the Wormhole's Appearance

Just as I had done for Gargantua (Chapter 8), I used Einstein's relativistic laws to deduce equations for the paths of light rays around and through the wormhole, and I worked out a procedure for manipulating my equations to compute the wormhole's gravitational lensing and thence what a camera sees when it orbits the wormhole or travels through it. After checking that my equations and procedure produced the kinds of images I expected, I sent them to Oliver and he wrote computer code capable of creating the quality IMAX images needed for the movie. Eugénie von Tunzelmann added background star fields and images of astronomical objects for the wormhole to lens, and then she, Oliver, and Paul began exploring the influence of my handles. Independently, I did my own explorations.

Eugénie kindly provided the pictures in Figures 15.2 and 15.4 for this book, in which we look at Saturn through the wormhole. (The resolutions of her pictures are far higher than my own crude computer code can produce.)

2 Most of the lensing occurs in the region where the wormhole's shape in the bulk is strongly curved. This is the region where its outward slope is steeper than 45 degrees, so I define the lensing width to be the radial distance, in the bulk, from the wormhole throat to the location with 45-degree outward slope (Figure 15.1).

The Wormhole's Length

We first explored the influence of the wormhole's length, with modest lensing (small lensing width); see Figure 15.2.

When the wormhole is short (top picture), the camera sees one distorted image of Saturn through the wormhole, the primary image, filling the right half of the wormhole's crystal-ball-like mouth. There is an extremely thin secondary, lenticular image on the left edge of the crystal ball. (The lenticular structure at the lower right is not Saturn; it is a distorted part of the external universe.)

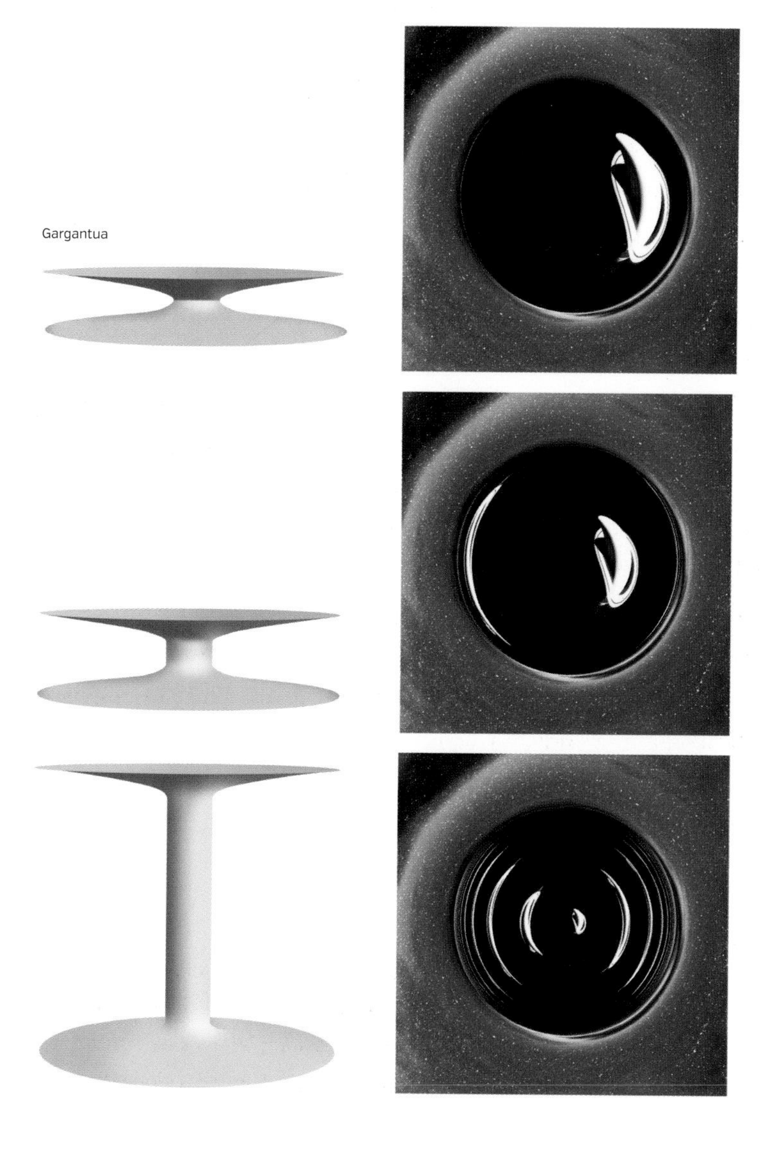

Fig. 15.2. *Left*: The wormhole, with small lensing width (just 5 percent of the wormhole's radius), viewed from the bulk. *Right*: What the camera sees. *Top to bottom*: Increasing wormhole length: 0.01, 1, and 10 times the wormhole's radius. *[From simulations by Eugénie von Tunzelmann's team using Oliver James' code based on my equations.]*

As the wormhole is lengthened (middle picture), the primary image shrinks and moves inward, the secondary image also moves inward, and a very thin lenticular tertiary image emerges from the right edge of the crystal ball.

With further lengthening (bottom picture), the primary image shrinks further, all the images move inward, a fourth image emerges from the left edge of the crystal ball, a fifth from the right, and so forth.

These behaviors can be understood by drawing light rays on the wormhole as seen from the bulk (Figure 15.3). The primary image is carried by the black light ray (1), which travels on the shortest possible path from Saturn to the camera, and by a bundle of rays tightly surrounding it. The secondary image is carried by a bundle surrounding the red ray (2), which travels around the wormhole's wall in the opposite direction to the black ray: counterclockwise. This red ray is the shortest possible counterclockwise ray from Saturn to the camera. The tertiary image is carried by a bundle surrounding the green ray (3), which is the shortest possible clockwise ray that makes more than one full trip around the wormhole. And the fourth image is carried by a bundle surrounding the brown ray (4): the shortest possible counterclockwise ray that makes more than one full trip around the wormhole.

Fig. 15.3. Light rays from Saturn, through the wormhole, to the camera.

Can you explain the fifth and sixth images? and explain why the images shrink when the wormhole is lengthened? and explain why the images appear to emerge from the edge of the wormhole's crystal-ball mouth and move inward?

The Wormhole's Lensing Width

Having understood how the wormhole's length affects what the camera sees, we then fixed the length to be fairly short, the same as the wormhole's radius, and varied the gravitational lensing. We increased the wormhole's lensing width from near zero to about half the wormhole's diameter and computed what that did to the images the camera sees. Figure 15.4 shows the two extremes.

With very small lensing width, the wormhole shape (upper left) has a sharp transition from the external universe (horizontal sheets) to the wormhole throat (vertical cylinder). As seen by the camera (upper right), the wormhole distorts the star field and a dark cloud in the upper left only slightly, near the wormhole's edge. Otherwise it simply masks the star field out, as would any opaque body with weak gravity, for example a planet or a spacecraft.

In the lower half of Figure 15.4, the lensing width is about half the wormhole's radius, so there is a slow transition from the throat (vertical cylinder) to the external universe (asymptotically horizontal sheet).

With this large lensing width, the wormhole strongly distorts the star field and dark cloud (lower right in Figure 15.3) in nearly the same way as does a nonspinning black hole (Figures 8.3 and 8.4), producing multiple images. And the lensing also enlarges the secondary and tertiary images of Saturn. The wormhole looks bigger in the lower half of Figure 15.3 than in the upper half. It subtends a larger angle as seen

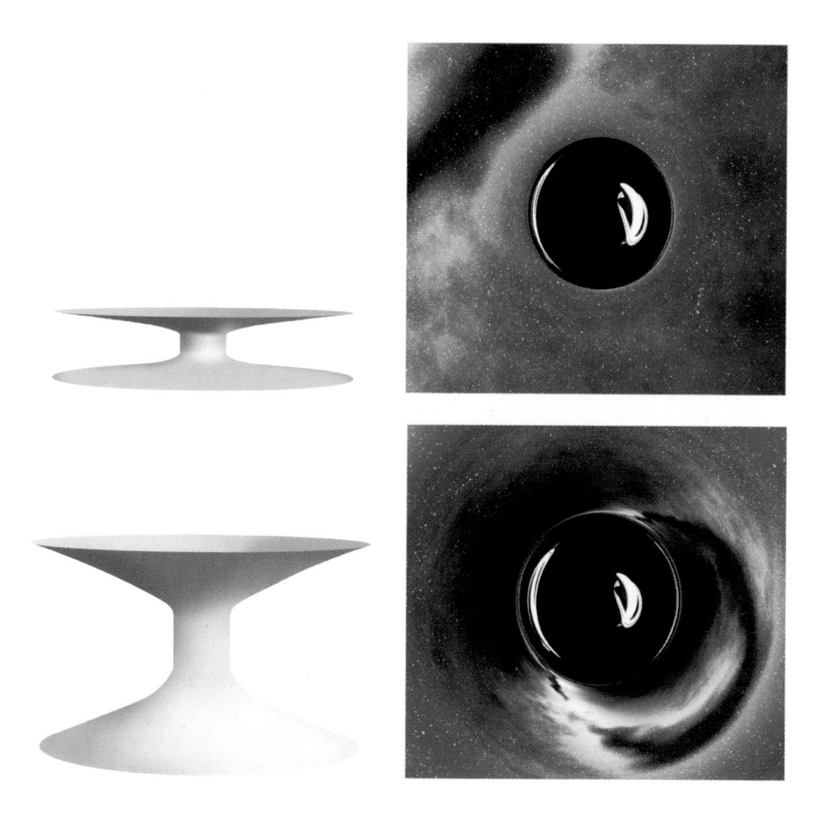

Fig. 15.4. Wormhole's gravitational lensing of a star field and Saturn: influence of the lensing width which is 0.014 and 0.43 times the wormhole radius in the top and bottom images. *[From simulations by Eugénie von Tunzelmann's team using Oliver James' code based on my equations.]*

by the camera. This is not because the camera is closer to the mouth; it is *not* closer. The camera is the same distance in both pictures. The enlargement is entirely due to the gravitational lensing.

Interstellar's Wormhole

When Chris saw the various possibilities, with varying wormhole length and lensing width, his choice was unequivocal. For medium and large length the multiple images seen through the wormhole would be confusing to a mass audience, so he made *Interstellar*'s wormhole very short: 1 percent of the wormhole radius. And he gave *Interstellar*'s wormhole a modest lensing width, about 5 percent of the wormhole radius, so the lensing of stars around it would be noticeable and intriguing, but much smaller than Gargantua's lensing.

The resulting wormhole is the one at the top of Figure 15.2. And in *Interstellar,* after the Double Negative team had created for its far side a galaxy with beautiful nebulae, dust lanes, and star fields, it is marvelous to behold (Figure 15.5). To me it is one of the movie's grandest sights.

The Trip Through the Wormhole

On April 10, 2014, I got an urgent phone call. Chris was having trouble with visualizing the *Endurance*'s trip through the wormhole and he wanted advice. I drove over to his Syncopy compound where post-production editing was underway, and Chris showed me the problem.

Using my equations, Paul's team had produced visualizations for wormhole trips with various wormhole lengths and lensing widths. For the short, modest-lensing wormhole depicted in the movie, the trip was quick and uninteresting. For a long wormhole, it looked like traveling through a long tunnel with walls whizzing past, too much like things we've seen in movies before. Chris showed me many variants with many bells and whistles, and I had to agree that none had the compelling freshness that he wanted. After sleeping on it, I still had no magic-bullet solution.

The next day Chris flew to London and closeted himself with Paul's

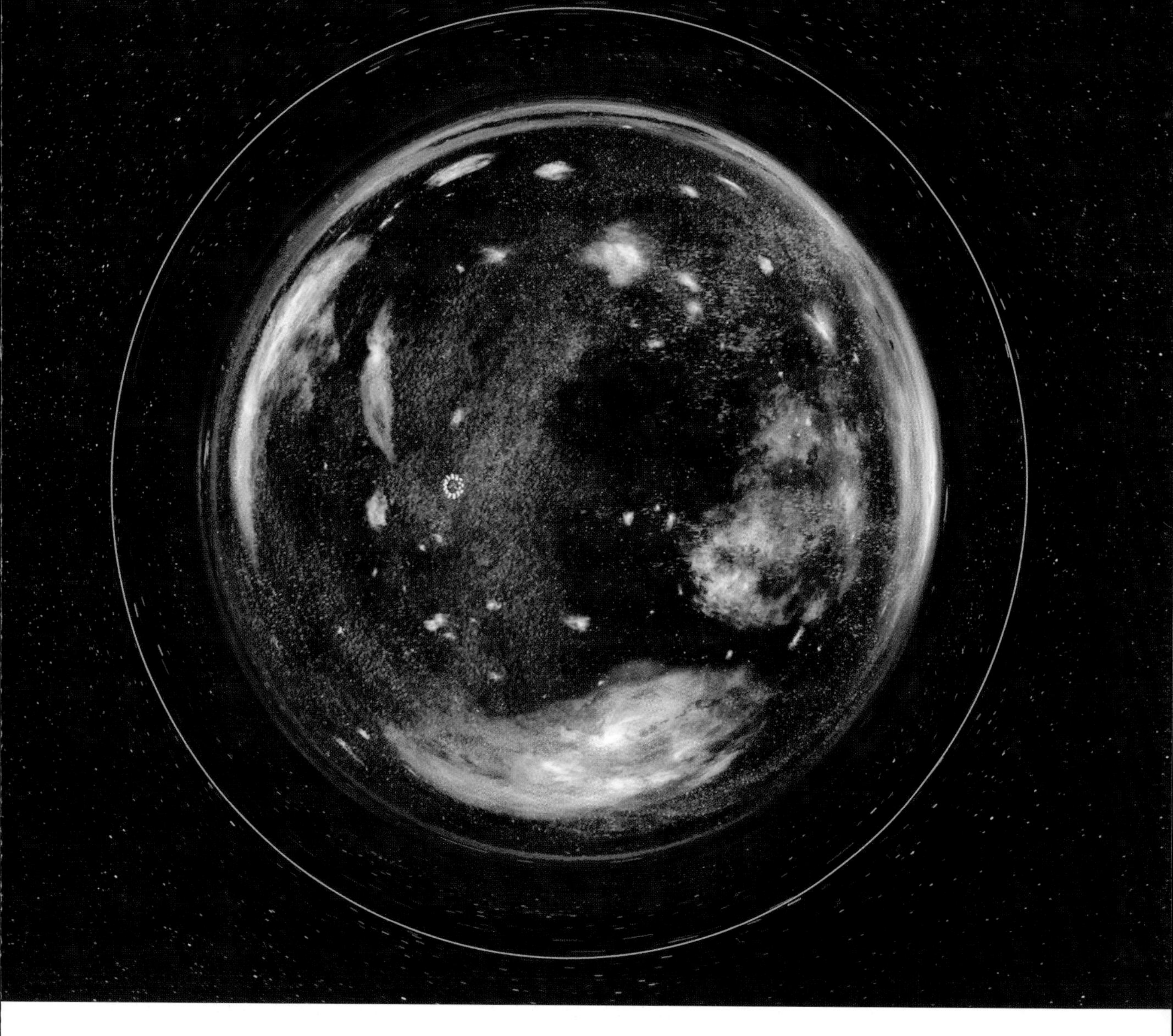

Double Negative team, searching for a solution. In the end, they were forced to abandon my wormhole equations and, in Paul's words, "go for a much more abstract interpretation of the wormhole's interior," an interpretation informed by simulations with my equations, but altered significantly to add artistic freshness.

When I experienced the wormhole trip in an early screening of *Interstellar,* I was pleased. Though not fully accurate, it captures the spirit and much of the feel of a real wormhole trip, and it's fresh and compelling.

What did you think?

Fig. 15.5. The wormhole as seen in a trailer for *Interstellar.* The *Endurance* is in front of the wormhole, near the center. Around the wormhole in pink I have drawn the Einstein ring, like that in Figure 8.4 for a non-spinning black hole. Primary and secondary images of gravitationally lensed stars move in the same way here as there. Looking at the trailer, can you identify some and trace their motion? *[From* Interstellar, *used courtesy of Warner Bros. Entertainment Inc.]*

16

Discovering the Wormhole: Gravitational Waves

Ⓣ

How might humans have discovered *Interstellar*'s wormhole? As a physicist, I have a favorite way. I describe it here in an extrapolation of *Interstellar*'s story—an extrapolation that, of course, is my own and not Christopher Nolan's.

LIGO Detects a Burst of Gravitational Waves

I imagine that decades before the movie begins, when Professor Brand was in his twenties, he was deputy director of a project called LIGO: The Laser Interferometer Gravitational Wave Observatory (Figure 16.1). LIGO was searching for ripples in the shape of space arriving at Earth from the distant universe. These ripples, called gravitational waves, are produced when black holes collide with each other, when a black hole tears a neutron star apart, when the universe was born, and in many other ways.

One day in 2019, LIGO was hit by a burst of gravitational waves far stronger than any ever before seen (Figure 16.2). The waves oscillated with an amplitude that grew and fell several times, and then cut off suddenly. The entire burst lasted for only a few seconds.

Fig. 16.1. *Top*: Aerial photograph of the LIGO gravitational wave detector at Hanford, Washington. *Left*: The LIGO control room where the detector is controlled and its signals are monitored.

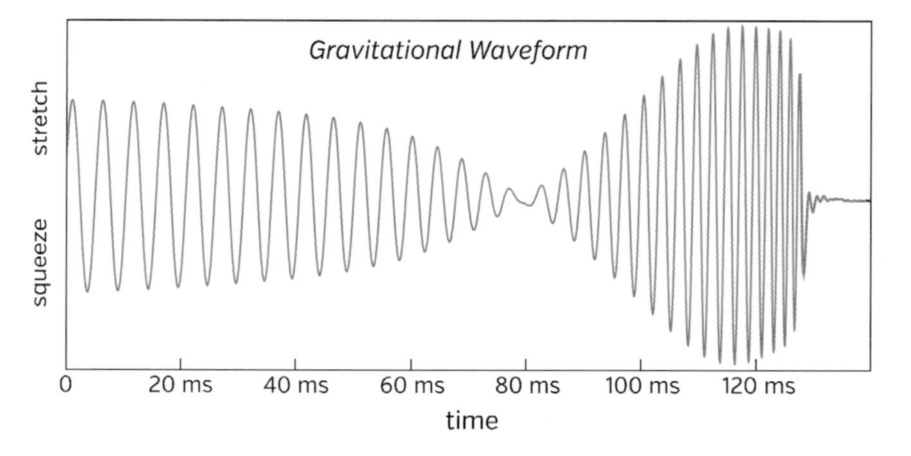

Fig. 16.2. The last 120 milliseconds (ms) of the gravitational waveform discovered by LIGO. *[Drawing by Kip based on simulations by Yanbei Chen and by Foucart et al. (2011).]*

By comparing the waves' shape (their "waveform"; Figure 16.2) with simulations performed on supercomputers, Professor Brand and his team deduced their source.

Neutron Star Orbiting a Black Hole

A neutron star, orbiting around a black hole, had emitted the waves. The star weighed 1.5 times as much as the Sun, the hole weighed 4.5 times the Sun, and the hole was spinning rapidly. The spin dragged space into motion, and the space whirl grabbed the star's orbit, forcing it to precess slowly, like a tilted top. The precession modulated the waves, causing them to rise and fall in amplitude (Figure 16.2).

The waves traveled out through the universe, carrying away energy (Figure 16.3). With its energy gradually decreasing, the star gradually spiraled inward toward the black hole. When the distance between the star and the hole had shrunk to 30 kilometers, the hole's tidal gravity began tearing the star apart. Ninety-seven percent of the stellar debris was swallowed by the black hole, and 3 percent was thrown outward, forming a tail of hot gas that the hole then sucked back inward to form an accretion disk.

Figure 16.4 shows a computer simulation of the last few millisec-

Fig. 16.3. Gravitational waves flowing out from the orbiting star and hole, as seen from the bulk. *[Drawing by LIGO Laboratory artist based on my hand sketch.]*

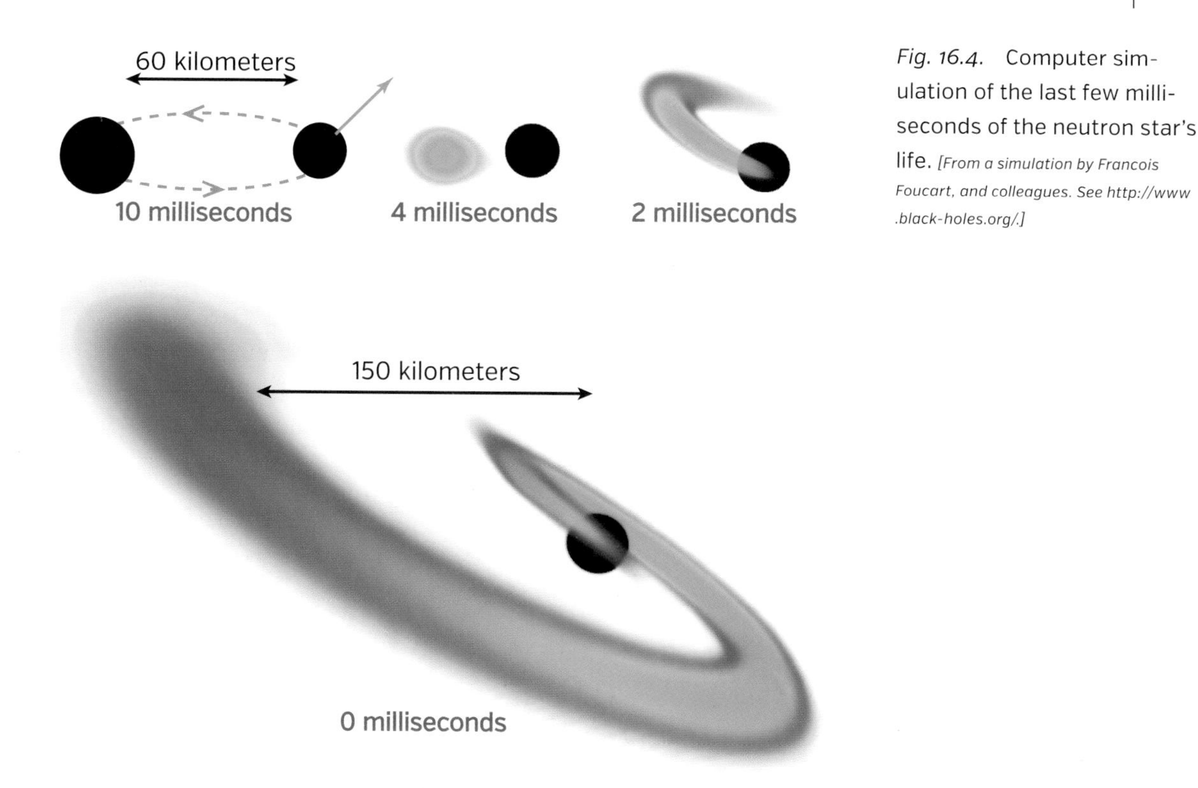

Fig. 16.4. Computer simulation of the last few milliseconds of the neutron star's life. *[From a simulation by Francois Foucart, and colleagues. See http://www.black-holes.org/.]*

onds of the star's life. At ten milliseconds before the end, the black hole is spinning around the red-arrowed axis, and the star is orbiting around the picture's vertical axis. At four milliseconds the hole's tendex lines are stretching the star apart. At two milliseconds, the hole's whirling space has thrown the stellar debris up into the hole's equatorial plane. At zero milliseconds, the debris is beginning to form an accretion disk.

Discovering the Wormhole

Looking back through LIGO's data for the preceding two years, Professor Brand and his team discovered very weak waves emitted by the neutron star. The star had a tiny mountain, a centimeter high and a few kilometers wide (such mountains are thought likely). As the mountain was carried around and around by the star's rotation, it produced waves that oscillated weakly but steadily, day after day after day.

By analyzing these steady waves with care, Professor Brand learned the direction to their source. The direction was unbelievable! The waves were coming from something in orbit around Saturn. As the Earth and Saturn moved in their orbits, the source was always near Saturn!

A neutron star orbiting Saturn? Impossible! A black hole accompanying the neutron star, with both orbiting Saturn? Even more impossible! Saturn would long ago have been destroyed, and the star's and hole's gravity would long ago have disrupted the orbits of all the Sun's planets, including Earth. With disrupted orbit, the Earth would have been carried close to the Sun and then far away. We would have been fried, frozen, and killed.

But there the waves were. Unequivocally emerging from near Saturn.

Professor Brand could find only one explanation: The waves must emerge from a wormhole that orbits Saturn. And their source, the black hole and neutron star, must be on the other end of the wormhole (Figure 16.5). The waves traveled outward from the star and hole. Small portions of the waves were captured by the wormhole, traveled through it, and then spread outward through the solar system with a small portion reaching Earth and passing through the LIGO gravitational wave detector.

Origin of This Story

A brief variant of this story was in the original 2006 treatment for *Interstellar* that Lynda Obst and I wrote. However, gravitational waves did not play a significant role in the rest of our treatment, nor in the subsequent screenplay that Jonathan Nolan wrote and Chris rewrote.

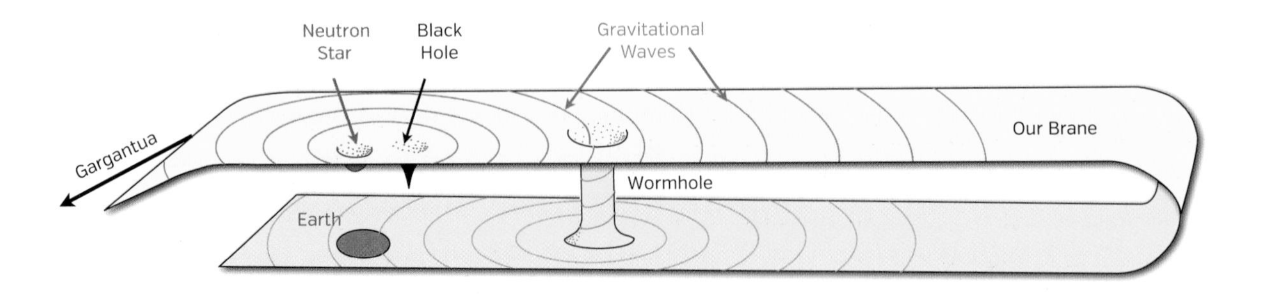

Fig. 16.5. Gravitational waves travel through the wormhole to Earth.

And even without gravitational waves, the amount of serious science in the movie was enormous. So when Chris sought ways to simplify *Interstellar*'s rich panoply of science, gravitational waves were a natural candidate for the ax. He jettisoned them.

For me, personally, Chris's decision was painful. I cofounded the LIGO Project in 1983 (together with Rainer Weiss at MIT and Ronald Drever at Caltech). I formulated LIGO's scientific vision, and I spent two decades working hard to help make it a reality. And LIGO today is nearing maturity, with the first detection of gravitational waves expected in this decade.

But Chris's reasons to jettison gravitational waves were compelling, so I didn't utter even a whisper of protest.

Gravitational Waves and Their Detectors

I indulge myself and tell you a bit more about gravitational waves before moving back to *Interstellar*.

Figure 16.6 is an artist's conception of some tendex lines emerging from two black holes that orbit each other counterclockwise, and collide. Recall that tendex lines produce tidal gravity (Chapter 4). The lines emerging from the holes' ends stretch everything they encounter, including the artist's friend, whom she has placed there. The lines emerging from the collision region squeeze everything they encounter. As the holes orbit around each other, they drag their tendex lines around, splaying outward and backward, like water from a whirling sprinkler.

The holes merge to form a single, larger black hole that is deformed and spinning counterclockwise, and that drags its tendex lines around and around. The tendex lines travel outward, like water from the sprin-

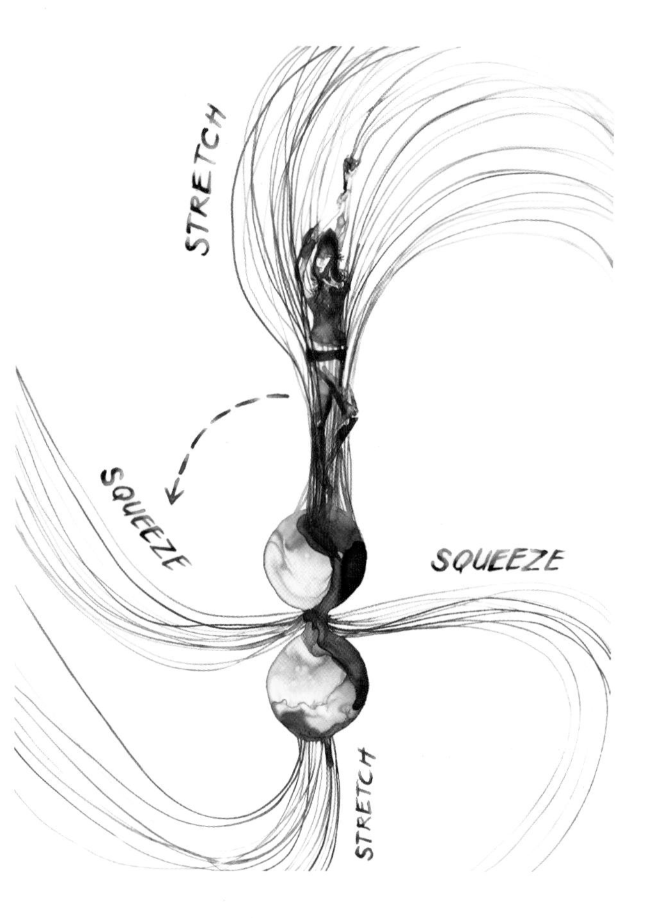

Fig. 16.6. Tendex lines from two black holes that collide while orbiting each other counterclockwise. *[Painting by Lia Halloran.]*

kler, creating the intricate pattern that I show in Figure 16.7. The red lines stretch. The blue lines squeeze.

A person at rest far from the hole experiences an oscillating stretch then squeeze then stretch as the tendex lines travel outward through her. The tendex lines have become a gravitational wave. Wherever the lines in the plane of the picture are strongly blue (strongly squeezing), there are strongly red lines coming out of the picture, that stretch. And wherever the lines in the picture are strongly red (stretching), there are blue (squeezing) lines pointing in the third direction, out of the picture. As the waves flow outward, the hole's deformation gradually grows weaker and the waves weaken.

When these waves reach Earth, they have the form that I show in the upper part of Figure 16.8. They stretch along one direction and squeeze along the other. The stretch and squeeze oscillate (from red right-left to blue right-left to red right-left, etc.) as the waves pass through the detector in the bottom part of Figure 16.8.

The detector consists of four huge mirrors (40 kilograms, 34 centimeters in diameter) that hang from overhead supports at the ends of two perpendicular arms. The waves' tendex lines stretch one arm

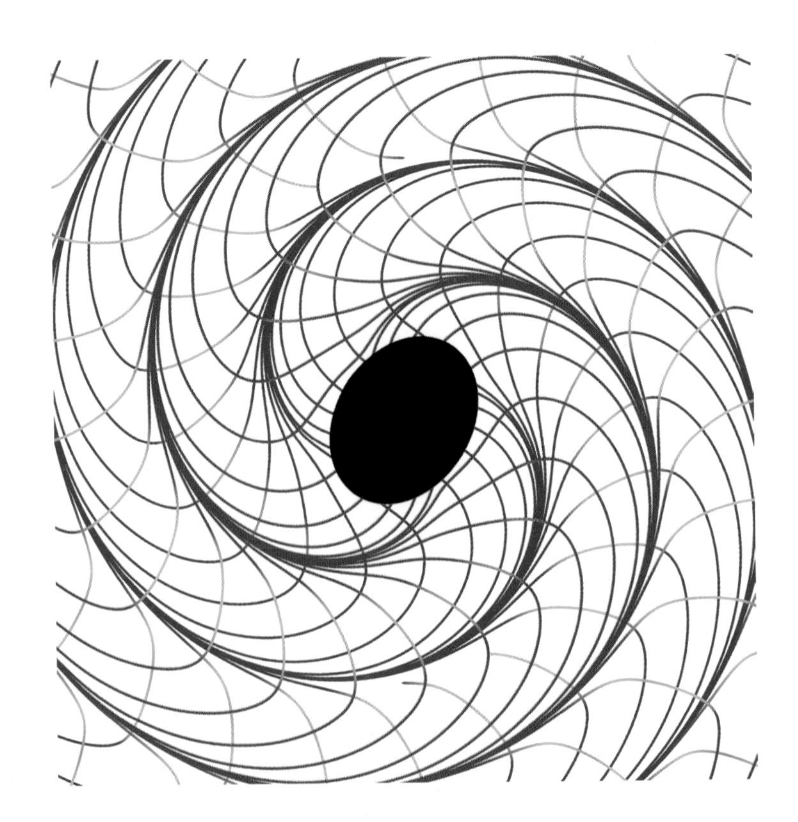

Fig. 16.7. Tendex lines from a spinning, deformed black hole. *[Drawing by Rob Owen.]*

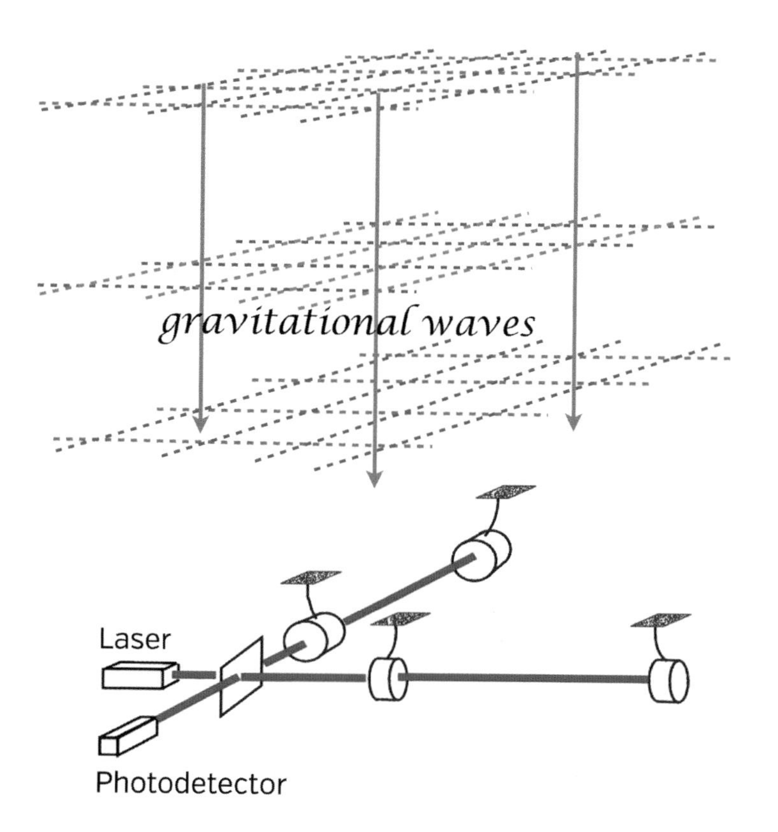

Fig. 16.8. Gravitational waves impinging on a LIGO detector.

while squeezing the other, and then squeeze the first while stretching the second, over and over and over again. The oscillating separation between mirrors is monitored with laser beams, by a technique called interferometry. Hence LIGO's name: Laser Interferometer Gravitational Wave Observatory.

LIGO is now an international collaboration of 900 scientists in seventeen nations, headquartered at Caltech. It is currently led by David Reitze (director), Albert Lazzarini (deputy director) and Gabriella Gonzalez (spokesperson for the collaboration). And in view of its huge potential payoffs for our understanding of the universe, it is funded primarily by US taxpayers, through the National Science Foundation.

LIGO has gravitational wave detectors in Hanford, Washington, and Livingston, Louisiana, and is planning to place a third in India. Scientists in Italy, France, and the Netherlands have built a similar interferometer near Pisa, and Japanese physicists are building one in a tunnel under a mountain. These detectors will all operate together, forming a giant worldwide network to explore the universe using gravitational waves.

Having trained many scientists who work on LIGO, in 2000 I

turned my own research in other directions. But I watch eagerly as LIGO and its international partners near maturity and near their first detections of gravitational waves.

The Warped Side of the Universe

Interstellar is an adventure in which humans encounter black holes, wormholes, singularities, gravitational anomalies, and higher dimensions. All these phenomena are "made from" warped space and time, or are tied intimately to that warping. This is why I like to call them the "warped side of the universe."

We humans, as yet, have very little experimental or observational data from the universe's warped side. That's why gravitational waves are important: they are made from warped space, and so they are the ideal tool for probing the warped side.

Suppose you had only seen the ocean on a very calm day. You would know nothing of the heaving seas and breaking ocean waves that come with a huge storm.

That is similar to our knowledge, today, of warped space and time. We know little about how warped space and warped time behave in a "storm"—when the shape of space is oscillating wildly and the rate of flow of time is oscillating wildly. For me this is a fascinating frontier of knowledge. John Wheeler, the creative coiner we met in earlier chapters, dubbed this "geometrodynamics": the wildly dynamical behavior of the geometry of space and time.

In the early 1960s, when I was Wheeler's student, he exhorted me and others to explore geometrodynamics in our research. We tried, and failed miserably. We didn't know how to solve Einstein's equations well enough to learn their predictions, and we had no way to observe geometrodynamics in the astronomical universe.

I've devoted much of my career to changing this. I cofounded LIGO with the goal of observing geometrodynamics in the distant universe. In 2000, when I turned my LIGO roles over to others, I cofounded a research group at Caltech aimed at simulating geometrodynamics on supercomputers, by solving Einstein's relativistic equations numerically. We call this project SXS: Simulating eXtreme Spacetimes. It is a collaboration with Saul Teukolsky's research group at Cornell University, and others.

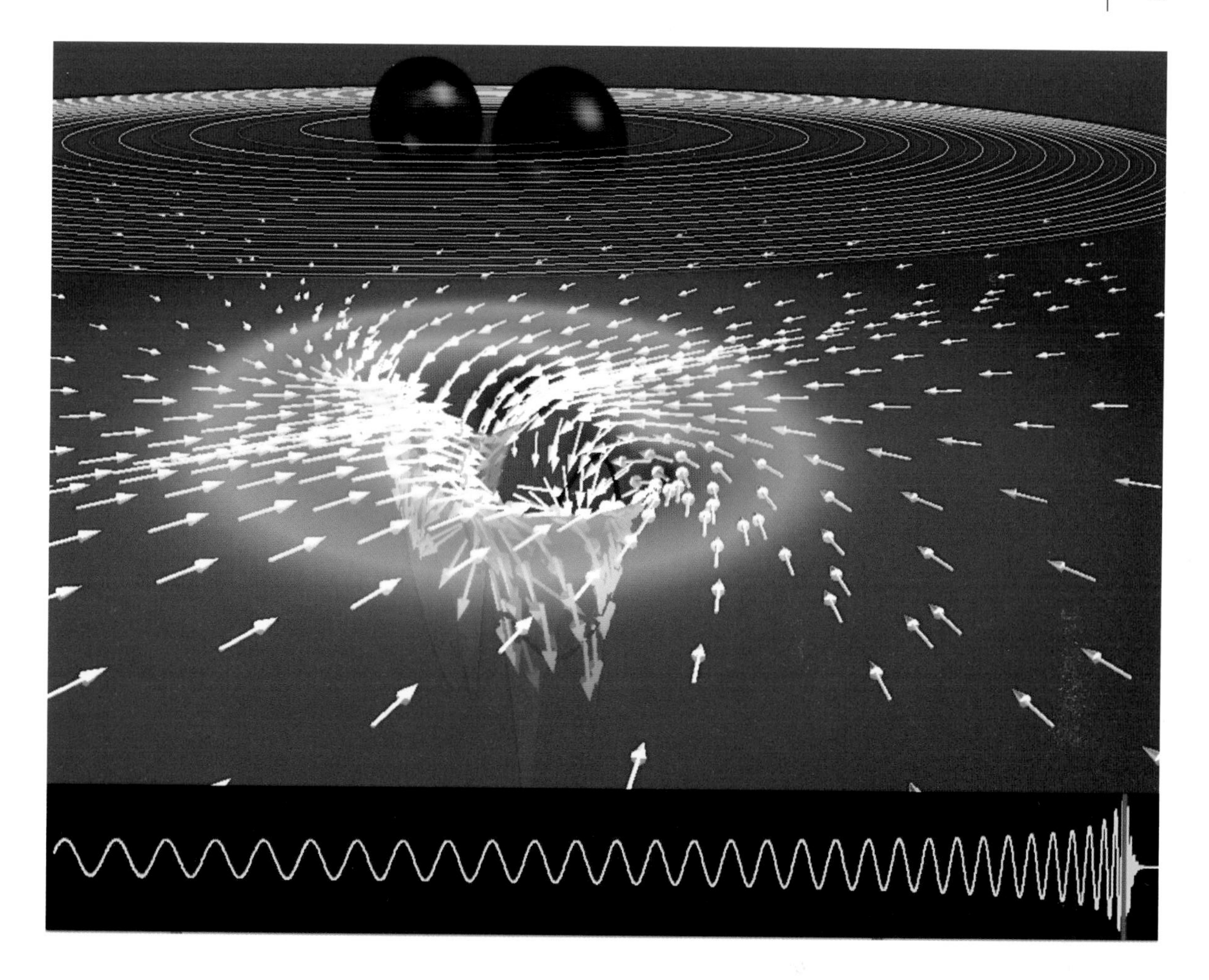

A wonderful venue for geometrodynamics is the collision of two black holes. When they collide, the holes set space and time into wild gyrations. Our SXS simulations have now reached maturity, and are beginning to reveal relativity's predictions (Figure 16.9). LIGO and its partners will observe the gravitational waves from colliding black holes within the next few years, and test our simulations' predictions. It's a wonderful era for probing geometrodynamics!

Fig. 16.9. Simulation of two black holes at their moment of collision. *Top*: The holes' orbits and shadows as seen in our universe. *Middle*: The holes' warped space and time as seen from the bulk, with arrows showing the dragging of space into motion and colors the warping of time. *Bottom*: The emitted gravitational waveform. This simulation is for identical, nonspinning black holes. *[From a movie by Harald Pfeifer of a simulation by the SXS team.]*

Gravitational Waves from the Big Bang

In 1975 Leonid Grishchuk, a dear Russian friend of mine, made a startling prediction: A rich plethora of gravitational waves was produced in the big bang, he predicted, by a previously unknown mechanism: Quantum fluctuations of gravity coming off the big bang were amplified enormously, he told us, by the universe's initial expansion; and when amplified, they became primordial gravitational waves. If discovered, these gravitational waves could bring us a glimpse of our universe's birth.

In subsequent years, as our understanding of the big bang matured, it became evident that the waves would be strongest at wavelengths nearly as large as the visible universe itself, billions of light-years' wavelength, and would likely be too weak for detection at LIGO's far shorter wavelengths, hundreds and thousands of kilometers.

In the early 1990s several cosmologists realized that these billion-light-years-long *gravitational* waves should have placed a unique imprint on *electromagnetic* waves that fill the universe, the so-called cosmic microwave background or CMB. A holy grail quickly emerged: search for that CMB imprint, from it infer the properties of the primordial gravitational waves that produced the imprint, and thereby explore the birth of the universe.

In March 2014, while I was writing this book, the CMB imprint was discovered by a team assembled by Jamie Bock (Figure 16.10),[1] a cosmologist down the hall from me at Caltech.

Fig. 16.10. The Bicep2 instrument, built by Jamie Bock's team, that discovered the imprint of primordial gravitational waves. Bicep2, at the South Pole, is here shown at twilight, which occurs only twice a year at the South Pole. It is surrounded by a shield to protect it from radiation from the surrounding ice sheet. The upper right inset shows the measured imprint on the CMB: a polarization pattern. The CMB's electric field points along the short dashed directions.

1 The formal leaders of the discovery team were Jamie and his former postdoctoral students John Kovac (now at Harvard) and Chao-Lin Kuo (now at Stanford), along with Clem Pryke (now at the University of Minnesota).

It was a fantastic discovery, but with a cautionary note: the imprint that Jamie and his team found might possibly be due to something else and not gravitational waves. As this book goes to press, intense efforts are underway to find out for sure.

If the imprint is really due to gravitational waves from the big bang, then this is the type of cosmological discovery that comes along perhaps once every fifty years. It brings us a glimpse of the universe a trillionth of a trillionth of a trillionth of a second after the universe's birth. It confirms theorists' prediction that the expansion of the universe at that early moment was exceedingly fast, "inflationarily fast" in cosmologists' jargon. It ushers in a whole new era for cosmology.

Having indulged my passion for gravitational waves, having seen how they could be used to discover *Interstellar*'s wormhole—and having explored the properties of wormholes, especially *Interstellar*'s—I now take you on a tour of the other side of the *Interstellar* wormhole. A tour of Miller's planet, Mann's planet, and the *Endurance*, which carries Cooper there.

V

EXPLORING GARGANTUA'S ENVIRONS

17

Miller's Planet

Ⓣ

The first planet that Cooper and his crew visit is Miller's. The most impressive things about this planet are the extreme slowing of time there, gigantic water waves, and huge tidal gravity. All three are related, and arise from the planet's closeness to Gargantua.

The Planet's Orbit

In my interpretation of *Interstellar*'s science, Miller's planet is at the blue location in Figure 17.1, very close to Gargantua's horizon. (See Chapters 6 and 7.)

Space there is warped like the surface of a cylinder. In the figure, the cylinder's cross sections are circles whose circumferences don't change as we move nearer to or farther from Gargantua. In reality, when we restore the missing dimension, the cross sections are spheroids, whose circumferences don't change as we move nearer or farther.

So why is this location different from any other on the cylinder? What makes this location special?

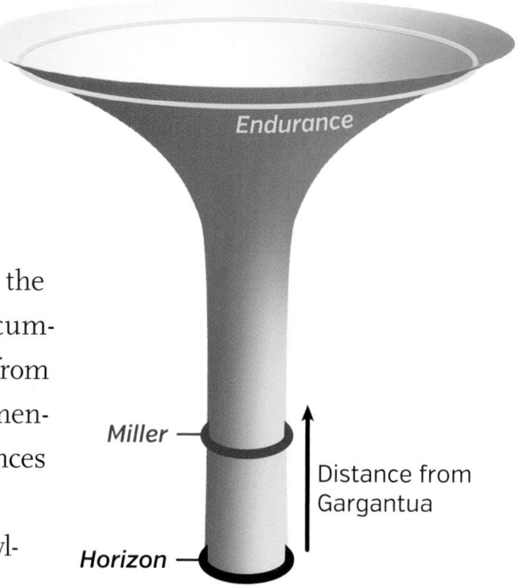

Fig. 17.1. The warped space around Gargantua as seen from the bulk, with one space dimension omitted. Also, the orbits of Miller's planet and the *Endurance*, parked and waiting for the crew to return.

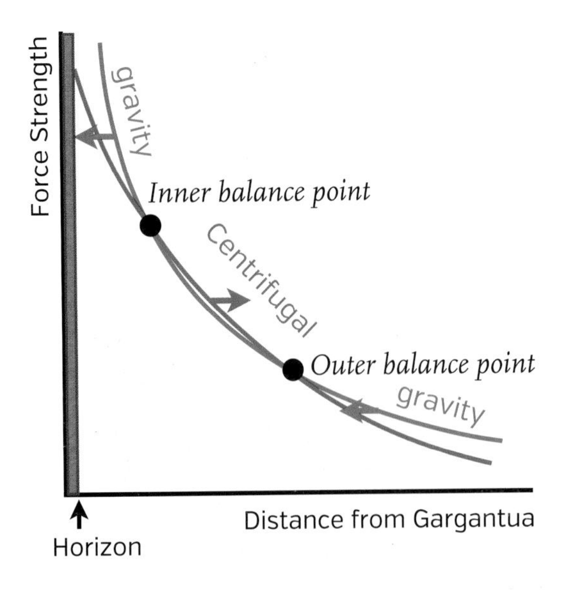

Fig. 17.2. The gravitational and centrifugal forces on Miller's planet.

The key to the answer is the warping of time, which does *not* show up in Figure 17.1. Time slows near Gargantua, and the slowing becomes more extreme as we get closer and closer to Gargantua's event horizon. Therefore, according to Einstein's law of time warps (Chapter 4), gravity becomes ultrastrong as we near the horizon. The red curve in Figure 17.2, which depicts the strength of the gravitational force, turns sharply upward. By contrast, the centrifugal force that the planet feels (the blue curve) has a more gradually changing slope. As a result, the two curves cross at two locations. There the planet can travel around Gargantua with the outward centrifugal force balancing the inward gravitational force.

At the inner balance point, the planet's orbit is unstable: If the planet gets pushed outward a tiny bit (for example, by the gravity of some passing comet), the centrifugal force wins the competition and pushes the planet further outward. If the planet is pushed inward, the gravitational force wins and the planet is pulled into Gargantua. This means Miller's planet can't live for long at the inner balance point.

The outer balance point, by contrast, is stable: If Miller's planet is there and gets pushed outward, gravity wins the competition and pulls the planet back in. If the planet gets pushed inward, centrifugal forces win and push it back out. So this is where Miller's planet lives, in my interpretation of *Interstellar.*[1]

1 The centrifugal force depends on the planet's orbital angular momentum, a measure of its orbital speed that is constant along its orbit (Chapter 10). In plotting how the force changes with distance from Gargantua in Figure 17.2, I hold that angular momentum constant. If the angular momentum were a bit smaller than the amount Miller's planet actually has, then the centrifugal force would everywhere be smaller, and the two curves in Figure 17.2 would not cross. There would be no balance point, and the planet would fall into Gargantua. That's why the location of Miller's planet in Figures 17.1 and 17.2 is the closest to Gargantua that the planet can stably live—the location I want, in order to get maximum slowing of time. For more details see *Some Technical Notes* at the end of this book.

The Slowing of Time, and Tidal Gravity

Among all stable, circular orbits around Gargantua, the orbit of Miller's planet is the closest to the black hole. This means it's the orbit with the maximum slowing of time. Seven years on Earth is one hour on Miller's planet. Time flows sixty thousand times more slowly there than on Earth! This is what Christopher Nolan wanted for his movie.

But being so close to Gargantua, in my interpretation of the movie, Miller's planet is subjected to enormous tidal gravity, so enormous that Gargantua's tidal forces almost tear the planet apart (Chapter 6). Almost, but not quite. Instead, they simply deform the planet. Deform it greatly (Figure 17.3). It bulges strongly toward and away from Gargantua.

Fig. 17.3. Tidal deformation of Miller's planet.

If Miller's planet were to rotate relative to Gargantua (if it didn't keep the same face toward Gargantua at all times), then as seen by the planet, the tidal forces would rotate. First the planet would be crushed east-west and stretched north-south. Then, after a quarter rotation, the crush would be north-south and the stretch east-west. These crushes and stretches would be enormous compared to the strength of the planet's mantle (its solid outer layers). The mantle would be pulverized, and then friction would heat it and melt it, making the whole planet red hot.

That's not at all what Miller's planet looks like! So the conclusion is clear: In my science interpretation, the planet must always keep the same face pointing toward Gargantua (Figure 17.4), or nearly so (as I discuss later).

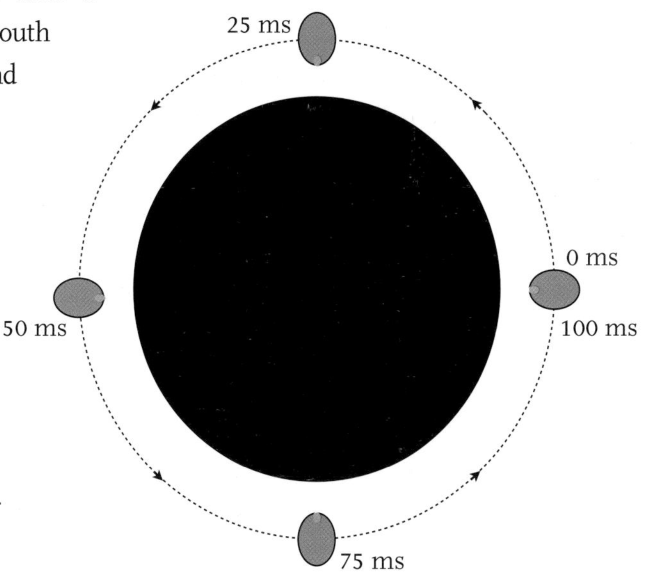

Fig. 17.4. The orbital motion and spin of Miller's planet relative to distant stars. The red spot on the planet's surface and the tidal bulge always face Gargantua.

The Whirl of Space

Einstein's laws dictate that, as seen from afar, for example, from Mann's planet, Miller's planet travels around Gargantua's billion-kilometer-circumference orbit once each 1.7 hours. This is roughly half the speed of light! Because of time's slowing, the Ranger's crew measure an orbital period sixty thousand times smaller than this: a tenth of a second. Ten trips around Gargantua per second. That's *really* fast! Isn't it far faster than light? No, because of the space whirl induced by Gargantua's fast spin. Relative to the whirling space at the planet's location, and using time as measured there, the planet is moving slower than light, and that's what counts. That's the sense in which the speed limit is enforced.

In my science interpretation of the movie, since the planet always keeps the same face pointed toward Gargantua (Figure 17.4), it must spin at the same rate as it orbits, ten revolutions per second. How can it possibly spin so fast? Won't centrifugal forces tear it apart? No; and again the savior is the whirl of space. The planet would feel *no* disruptive centrifugal forces if it were spinning at precisely the same rate as space near it whirls, which it is almost doing! So centrifugal forces due to its rotation, in fact, are weak. If, instead, it were nonspinning relative to the distant stars, it would turn at ten revolutions per second relative to whirling space and so would be torn apart by centrifugal forces. It's weird what relativity can do.

Giant Waves on Miller's Planet

What could possibly produce the two gigantic water waves, 1.2 kilometers high, that bear down on the Ranger as it rests on Miller's planet (Figure 17.5)?

I searched for a while, did various calculations with the laws of physics, and found two possible answers for my science interpretation of the movie. Both answers require that the planet be *not quite* locked to Gargantua. Instead it must rock back and forth relative to Gargantua by a small amount, from the orientation on the left of Figure 17.6, to that on the right, then back to the left, and so on.

This rocking is a natural thing, as you can see by looking at Gargantua's tidal gravity.

In Figure 17.6, I describe the tidal gravity by *tendex lines* (Chapter 4). No matter which way the planet is tilted (left or right in Figure 17.6), Gargantua's blue squeezing tendex lines push its sides in, which

Fig. 17.5. A giant water wave bearing down on the Ranger. *[From* Interstellar, *used courtesy of Warner Bros. Entertainment Inc.]*

Fig. 17.6. The rocking of Miller's planet in response to Gargantua's tidal gravity: its stretching tendex lines (*red*) and squeezing tendex lines (*blue*).

drives the planet back toward its preferred orientation, the one with its bulges nearest Gargantua and farthest away (Figure 17.3). Similarly, Gargantua's red stretching tendex lines pull its bottom bulge toward Gargantua and push its top bulge away from Gargantua. This also drives the planet back toward its preferred orientation.

The result is a simple rocking of the planet, back and forth, if the tilts are small enough that the planet's mantle isn't pulverized. When

I computed the period of this rocking, how long it takes to swing from left to right and back again, I got a joyous answer. About an hour. The same as the observed time between giant waves, a time chosen by Chris without knowing about my science interpretation.

The first explanation for the giant waves, in my science interpretation, is a sloshing of the planet's oceans as the planet rocks under the influence of Gargantua's tidal gravity.

A similar sloshing, called "tidal bores," happens on Earth, on nearly flat rivers that empty into the sea. When the ocean tide rises, a wall of water can go rushing up the river; usually a tiny wall, but very occasionally respectably big. You can see an example in the top half of Figure 17.7: a tidal bore on the Qiantang River in Hangzhou, China, in August 2010. Though impressive, this tidal bore is very small compared to the 1.2-kilometer-high waves on Miller's planet. But the Moon's tidal gravity that drives this tidal bore is tiny—really tiny—compared to Gargantua's huge tidal gravity!

My second explanation is tsunamis. As Miller's planet rocks, Gargantua's tidal forces may not pulverize its crust, but they do deform the crust first this way and then that, once an hour, and those deformations could easily produce gigantic earthquakes (or "millerquakes," I suppose we should call them). And those millerquakes could generate tsunamis on the planet's oceans, far larger than any tsunami ever seen on Earth, such as the one that hit Miyako City, Japan, on March 11, 2011 (bottom half of Figure 17.7).

Past History of Miller's Planet

It is interesting to speculate about the past and future history of Miller's planet. Try it using as much physics as you know or can scrounge up from the web or elsewhere. (This is not easy!) Here are some things you might think about.

How old is Miller's planet? If, as an extreme hypothesis, it was born in its present orbit when its galaxy was very young (about 12 billion years ago), and Gargantua has had its same ultrafast spin ever since, then the planet's age is about 12 billion years divided by 60,000 (the slowing of time on the planet): 200,000 years. This is awfully young

Fig. 17.7. Top: A tidal bore on the Qiantang River. *Bottom*: A tsunami in Miyako City.

compared to most geological processes on Earth. Could Miller's planet be that young and look like it looks? Could the planet develop its oceans and oxygen-rich atmosphere that quickly? If not, how could the planet have formed elsewhere and gotten moved to this orbit, so close to Gargantua?

How long can the planet's rocking continue until friction inside the planet converts all the rocking energy to heat? And how long could it have rocked in the past? If a lot shorter than 200,000 years, then perhaps something set it rocking. What could have done so?

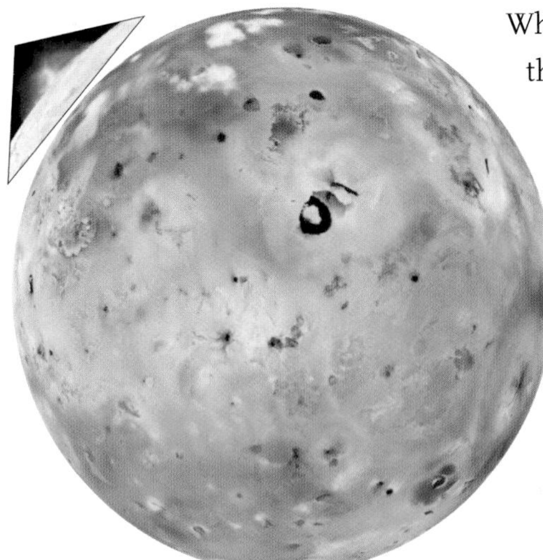

When friction converts rocking energy to heat, how hot does the planet's interior get? Hot enough to trigger volcanos and lava flows?

Jupiter's moon Io is a remarkable example of this. Io, the large moon that orbits closest to Jupiter's surface, doesn't rock. But it does move closer and farther from Jupiter along an elliptical orbit, so it feels Jupiter's tidal gravity strengthen then weaken then strengthen, much like Miller's planet feels Gargantua's tidal gravity oscillate. This heats Io enough to produce huge volcanos and lava flows (Figure 17.8).

Fig. 17.8. Io as photographed by the *Galileo* spacecraft shows many volcanos and lava flows. *Inset*: A 50-kilometer-high volcanic plume.

The Appearance of Gargantua from Miller's Planet

In *Interstellar*, as the Ranger approaches Miller's planet carrying Cooper and his crew, we see Gargantua in the sky above, 10 degrees across (twenty times larger than the Moon as seen from Earth!) and surrounded by its bright accretion disk. See Figure 17.9. As startlingly impressive as this may be, Gargantua's angular size has actually been reduced greatly from what it would really be at the location of Miller's planet.

If Miller's planet is, indeed, close enough to Gargantua to experience extreme time slowing—as I chose for my interpretation of the movie—then it must be deep into the cylindrical region of Gargantua's warped space, as depicted in Figure 17.1. It seems likely, then, that if you look down the cylinder from Miller's planet you will see Gargantua, and if you look up the cylinder you will see the external universe; so Gargantua should encompass roughly half of the sky (180 degrees) around the planet and the universe the other half. Indeed, that is what Einstein's relativistic laws predict.

It also seems clear that, since Miller's planet is the closest anything can live stably, without falling into Gargantua, the entire accretion disk should be outside the orbit of Miller's planet. Therefore, as the crew approach the planet, they should see a giant disk above them and a giant black-hole shadow below. Again, that is what Einstein's laws predict.

If Chris had followed these dictates of Einstein's laws, it would have

Fig. 17.9. Gargantua and its disk, partially eclipsed by Miller's planet, as the Ranger, in the foreground, descends toward landing. *[From* Interstellar, *used courtesy of Warner Bros. Entertainment Inc.]*

spoiled his movie. To see such fantastic sights so early in the movie would make the movie's climax, when Cooper falls into Gargantua, visually anticlimactic. So Chris consciously saved such sights for the end of the movie; and invoking artistic license, near Miller's planet he depicted Gargantua and its disk together, "just" twenty times bigger than the Moon looks from Earth.

Although I'm a scientist and aspire to science accuracy in science fiction, I can't blame Chris at all. I would have done the same, had I been making the decision. And you'd have thanked me for it.

18

Gargantua's Vibrations

Ⓣ

While Cooper and Amelia Brand are on Miller's planet, Romilly stays behind in the *Endurance,* observing Gargantua. From exquisitely accurate observations, he hopes to learn more about gravitational anomalies. Above all (I presume), he hopes that quantum data from Gargantua's singularity (Chapter 26) will leak out through the event horizon, bringing information about how to control the gravitational anomalies (Chapter 24). Or, in Romilly's pithy language, bringing information for "solving gravity."

When Amelia Brand returns from Miller's planet, Romilly tells her, "I learned what I could from studying the black hole, but I couldn't send anything to your father. We've been receiving but nothing gets out."

What did Romilly observe? He's not specific, but I presume he would focus on Gargantua's vibrations, and I offer this chapter's extrapolation of the movie for that.

Vibrations of Black Holes

In 1971 Bill Press, a student of mine at Caltech, discovered that black holes can vibrate at special, *resonant frequencies,* in much the same way as a violin string vibrates.

When a violin string is plucked just right, it emits a very pure tone: sound waves with a single frequency. When plucked a little differently, it emits that pure tone and also higher harmonics of the pure tone. In other words (if the string is firmly clamped, with the clamping finger not moving around) its vibrations produce sound at only a discrete set of frequencies, the string's resonant frequencies.

The same is true of a wine glass whose rim you rub with your finger, and a bell struck by a hammer. And also a black hole disturbed by something falling into it, Press discovered.

A year later Saul Teukolsky, another of my students, used Einstein's relativistic laws to work out a mathematical description of these resonant vibrations for a spinning black hole. (That's the best thing about teaching at Caltech; we get fabulous students!) By solving Teukolsky's equations, we physicists can compute a black hole's resonant frequencies. But solving them for an extremely fast spinning hole (like Gargantua) is very difficult. So difficult that it was not done successfully until forty years later—by a collaboration in which the lead players again were two Caltech students: Huan Yang and Aaron Zimmerman.

In September 2013, Ritchie Kremer, the property master for *Interstellar* (the person in charge of props) asked me for observational data that Romilly could show to Brand. Of course, I turned to the world's best experts for help: Yang and Zimmerman. They quickly produced tables of Gargantua's resonant vibration frequencies and also of the rates that the vibrations die out by feeding energy into gravitational waves—tables based on their own calculations using Teukolsky's equations. Then they added fake observational numbers to go along with the theoretical predictions and I added pictures of Gargantua's event horizon (or rather, the edge of its shadow), pictures from simulations by the *Interstellar* visual-effects team at Double Negative. The result was Romilly's observational data set.

When Christopher Nolan filmed the scene where Romilly discusses his observations with Amelia Brand, Romilly wound up not actually showing her his data set. It was there on a table, but he didn't pick it up. However, the data set is central to my science extrapolation of *Interstellar*.

Gargantua's Resonant Vibrations

Figure 18.1 is the data set's first page. Each line of data on that page refers to a single resonant frequency at which Gargantua vibrates.

The first column is a three-number code for the shape of Gargantua's vibrations and the picture is a still from a movie Romilly took, in my extrapolation of *Interstellar*, which verified that the vibrations had the predicted shape. The second column of data is the vibration frequency and the third is the rate at which this vibration dies out, as predicted by Teukolsky's equations.[1] The fourth and fifth columns show the difference between Romilly's observations and the theoretical predictions.

In my extrapolation Romilly finds a few anomalies, severe disagreements between his observations and the theory. He prints the disagreements in red. On page one of the data set (Figure 18.1), there is just one anomaly, but the disagreement is severe: thirty-nine times larger than the uncertainty in his measurements!

These anomalies might be helpful in "solving gravity" (learning how to harness the anomalies), Romilly thinks, in my extrapolation. He wishes he could transmit what he has learned to Professor Brand back on Earth, but the outbound communication link has been severed, so he's frustrated.

Even more, he wishes he could see inside Gargantua, to extract the crucial quantum data embedded in its singularity (Chapter 26). But he can't.

And he doesn't know whether the anomalies he observed are encoding some of the quantum data or not. Perhaps, with the hole spinning so rapidly, some of the quantum data leaked out through the horizon

1 The table's numerical values for the resonant frequencies are not in familiar units. To convert to familiar units, we must multiply by the cube of the speed of light and divide by $2\pi GM$, where $\pi = 3.14159...$, G is Newton's gravitational constant, and M is Gargantua's mass. This conversion factor is approximately one vibration per hour, so the first predicted frequency in the table is about 0.67 vibrations per hour. The conversion factor for the die-out rate is the same.

GARGANTUA

Quasinormal Mode Frequencies – Averaged Over All Data

DM Family

Mode	Theory		Observed/Theory - 1	
(l,m,n)	Re(ω) M	Im(ω) M	Re(ω) M	Im(ω) M
(2,1,0)	0.6664799	0.05541304	0.000054±23	0.00038±44
(2,1,1)	0.6665907	0.1662391	0.000008±8	0.00025±26
(2,1,2)	0.6667016	0.2770652	0.000040±17	0.00039±39
(2,1,3)	0.6668124	0.3878913	0.000016±24	0.00051±8
(2,1,4)	0.6669232	0.4987174	0.000003±25	0.00005±8
(2,0,0)	0.5235067	0.0809975	0.000057±10	0.00017±19
(2,0,1)	0.5236687	0.2429925	0.000029±9	0.00065±13
(2,0,2)	0.5238307	0.4049875	0.000005±31	0.00042±15
(2,0,3)	0.5239927	0.5669825	0.000023±12	0.00039±50
(2,0,4)	0.5241547	0.7289775	0.000041±61	0.00003±46
(3,2,0)	1.0749379	0.03192427	0.000014±91	0.00009±71
(3,2,1)	1.0750018	0.09577282	0.000019±32	0.00021±24
(3,2,2)	1.0750656	0.1596214	0.000004±25	0.00006±21
(3,2,3)	1.0751295	0.2234699	0.000024±14	0.0011±19
(3,2,4)	1.0751933	0.2873185	0.000032±38	0.00007±28
(3,1,0)	0.8623969	0.06574082	0.000004±74	0.00051±27
(3,1,1)	0.8625284	0.1972225	0.00039±1	0.00016±9
(3,1,2)	0.8626599	0.3287041	0.000019±35	0.00057±41
(3,1,3)	0.8627914	0.4601857	0.000030±35	0.00002±21

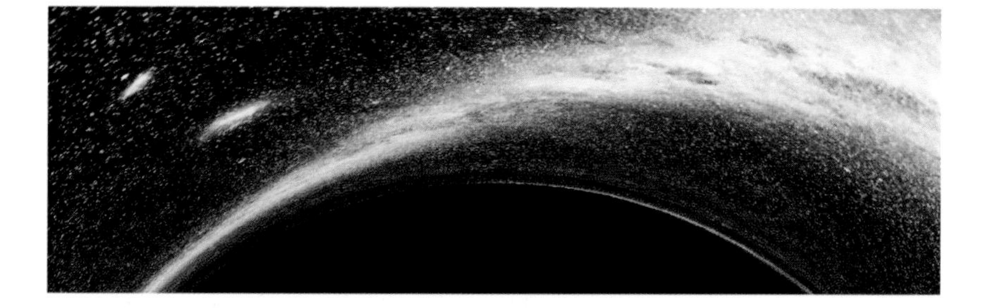

Fig. 18.1. The first page of the data that Yang and Zimmermann prepared for Romilly to show to Amelia Brand. *[Prop from* Interstellar, *used courtesy of Warner Bros. Entertainment Inc.]*

and produced the anomalies. Maybe Professor Brand could figure that out, if only Romilly could transmit the data to him.

I say a lot more later (Chapters 24–26) about gravitational anomalies, and quantum data from inside Gargantua as the key to harnessing the anomalies. But that's later. For now, let's continue our exploration of Gargantua's environs, turning next to Mann's planet.

Mann's Planet

Ⓣ

After discovering that Miller's planet is hopeless for human colonization, Cooper and his crew travel to Mann's planet.

The Planet's Orbit and Lack of Sun

I have deduced a plausible orbit for Mann's planet from two things in *Interstellar*:

First, Doyle says the trip to Mann's planet will require months. From this I infer that, when the *Endurance* arrives at Mann's planet, it must be far from Gargantua's vicinity where the trip began. Second, almost immediately after the *Endurance*'s explosive accident in orbit around Mann's planet, the crew find the *Endurance* being pulled toward Gargantua's horizon. From this I infer that, when they leave Mann's planet, the planet must be near Gargantua.

To achieve both requirements, the orbit of Mann's planet must be highly elongated. And to avoid the planet's being engulfed by Gargantua's accretion disk as it nears Gargantua, the orbit, so far as possible, must be far above or below Gargantua's equatorial plane, where the disk resides.

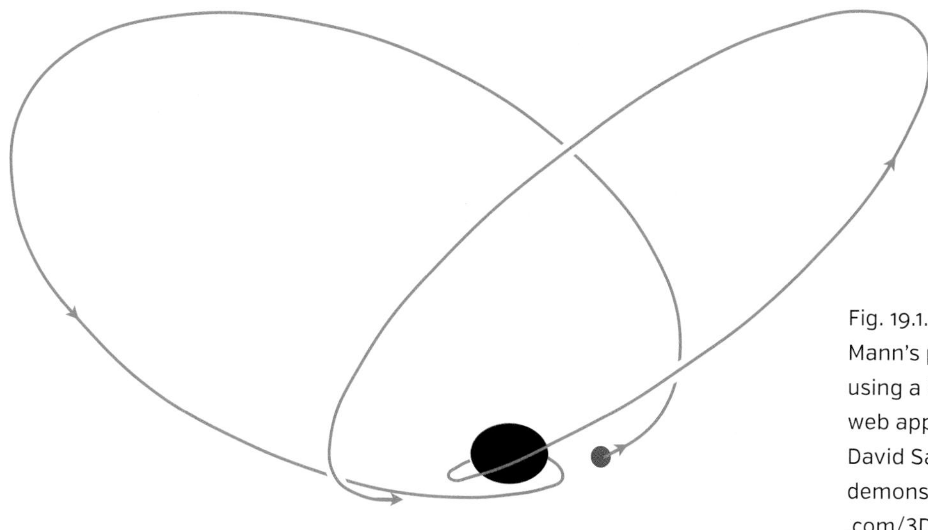

Fig. 19.1. A possible orbit for Mann's planet, computed using a highly user-friendly web application written by David Saroff; see http://demonstrations.wolfram.com/3DKerrBlackHoleOrbits.

This dictates an orbit something like that shown in Figure 19.1, though extending much farther from Gargantua, to 600 Gargantua radii or more.[1] Like the orbit of Halley's comet in our solar system (Figure 7.5), the planet swings close around Gargantua then flies out to a large distance, then returns, swings around Gargantua, and flies out again. The whirl of space near Gargantua makes the planet fly around Gargantua once or twice on each swing by, and makes its orbit precess through a large angle from one outward excursion to the next, as shown in the figure.

Mann's planet can't be accompanied by a sun on its inward and outward journeys because, when near Gargantua, huge tidal forces would pry the planet and its sun apart, sending them onward in markedly different orbits. Therefore, like Miller's planet, it must be heated and lit by Gargantua's anemic accretion disk.

1 In the movie, when the *Endurance* is in orbit around Mann's planet, we see Gargantua subtending about 0.9 degrees on the sky—nearly twice the size of the Moon as seen from Earth. From this I compute that Mann's planet is about 600 Gargantua radii from the black hole. At this distance, the time required for the planet to travel inward to near Gargantua is at least forty days—a lot longer than the crew *seem* to spend on and near Mann's planet, but reasonable for the outward trip to reach the planet; see Chapter 7.

The Trip to Mann's Planet

The *Endurance*'s trip to Mann's planet begins near Gargantua and ends far from it. Such a trip—in my scientist's interpretation of the movie—requires two gravitational slingshots (Chapter 7), one at the beginning of the trip and one at the end.

At the beginning, the challenges are twofold: In its parking orbit near Gargantua, the *Endurance* is moving at a third of light speed, $c/3$, in the wrong direction, a circumferential orbit around Gargantua; it must be deflected into radial motion, away from Gargantua. And the *Endurance* isn't moving fast enough. Gargantua's gravitational pull is so strong that, if the *Endurance* is deflected onto a radial trajectory but still has its starting speed of $c/3$, then Gargantua will pull it to a halt by the time it has traversed only a small fraction of the distance to Mann's planet. To overcome Gargantua's gravity and reach Mann's planet moving with the same speed as the planet, roughly $c/20$, the first slingshot must accelerate the *Endurance* up to nearly half the speed of light. To achieve this, Cooper must find an intermediate-mass black hole (IMBH) at an appropriate location and moving with a suitable velocity.

Finding the necessary IMBH is not easy, and having found it, reaching it at the right point and moment in its orbit may not be easy. Most of the months'-long trip may be spent reaching the IMBH, and it might entail considerable waiting for the IMBH to arrive. Once the slingshot is completed, the trip to Mann's planet, with speed about $c/2$ initially and gradual slowing to roughly $c/20$, will take roughly an additional forty days.

In the second slingshot, near Mann's planet, the *Endurance* swings around a suitable IMBH and soars into a gentle rendezvous with the planet: a rendezvous that doesn't require much rocket fuel.

Arrival at Mann's Planet: Ice Clouds

In the movie, the *Endurance* parks in orbit around Mann's planet, and then Cooper and his crew descend to the planet in a Ranger.

Fig. 19.2. The Ranger scraping the edge of an "ice cloud" on Mann's planet. *[From* Interstellar, *used courtesy of Warner Bros. Entertainment Inc.]*

The planet is covered with ice, as one might expect, since (in my interpretation) it spends most of its life far from the warmth of Gargantua's accretion disk. As the Ranger nears the planet, we see it maneuver among what appear to be clouds, but then it scrapes along one (Figure 19.2) and we discover the cloud is actually made from some sort of ice.

Motivated by a conversation with Paul Franklin, I imagine that these clouds are largely frozen carbon dioxide, "dry ice," and they are starting to be warmed as the planet is on its inward excursion toward the accretion disk, as in Figure 19.1. When warmed, dry ice sublimates—vaporizes—and so what appears to be clouds may be a mixture of dry ice and sublimating vapor; perhaps mostly vapor. At lower altitudes, where the Ranger lands, temperatures are higher and the ice on which they land is presumably all frozen water.

Dr. Mann's Geological Data

In the movie, Dr. Mann has been searching for organic material on his planet and he claims to have found promising evidence. Promising but not definitive. He shows his data to Brand and Romilly.

The data consist of field notes that indicate where each rock sample was collected and the geological environment there, together with chemical analyses of the sample. Those chemical analyses are Dr. Mann's evidence of organics.

Figure 19.3 shows a page from these data. The data were actually

Fig. 19.3. *Top:* Romilly (played by David Oyelowo) and Brand (played by Anne Hathaway) discuss Dr. Mann's geological data with him. *Bottom:* One page of data, prepared for the movie by Erika Swanson: the results of chemical analyses of rocks collected on the purported surface of the planet. Several rocks show promising evidence of organic material that could have arisen from living things. *[From* Interstellar*, used courtesy of Warner Bros. Entertainment Inc.]*

sample	cytochromes	Redox enzymes	PA hydrocarbons	formaldehyde
EBL-VR01	0.03	0.379	8.7	1.64
EBL-VR03	0.02	0.103	2.3	1.20
EBL-VR04	0.02	0.170	3.9	1.38
EBL-OS01	0.02	0.128	2.9	1.28
EBL-OS02	0.01	0.038	0.8	0.88
EBL-OS04	0.01	0.020	0.4	0.71
GBO-VR01	0.04	0.426	9.7	1.67
GBO-VR02	0.02	0.155	3.5	1.34
GBO-VR03	0.01	0.015	0.3	0.64
GBO-VR05	0.02	0.115	2.6	1.24
GBO-OS01	0.04	0.613	14.0	1.76
GBO-OS02	0.00	0.009	0.1	0.50
GBO-OS03	0.02	0.115	2.6	1.24
GBO-OS04	0.03	0.237	5.4	1.49
EFO-VR02	0.01	0.053	1.2	0.98
EFO-VR03	0.02	0.186	4.2	1.41
EFO-VR05	0.02	0.103	2.3	1.20
EFO-VR08	0.05	0.938	21.5	1.79
EFO-VR11	0.07	1.648	37.9	1.64
EFO-OS01	0.00	0.003	0.0	0.25
EFO-OS02	0.03	0.219	5.0	1.46
EFO-OS03	0.01	0.045	1.0	0.93
EFO-KS01	0.02	0.128	2.9	1.28

interesting

very promising!

prepared for the movie by Erika Swanson, a talented geology PhD student at Caltech. Erika has done fieldwork and chemical analyses somewhat similar to Dr. Mann's.

In the movie, it turns out that Dr. Mann has faked his data. That's a bit ironic since, of course, Erika faked her data too. She has never made a field trip to Mann's planet. Perhaps someday . . .

In this book I say nothing about the tragedy of Dr. Mann. It's a human tragedy, involving little science. The tragedy's climax is an explosion that severely damages the *Endurance*. The explosion, the damage, and the *Endurance*'s design: that's the stuff of science and engineering, so let's discuss them.

20

The *Endurance*

Ⓣ

Fig. 20.1. The *Endurance*, with two Rangers and two landers docked onto its central, control module. The Rangers are oriented out of the *Endurance*'s ring plane; the landers, parallel to it. *[From* Interstellar, *used courtesy of Warner Bros. Entertainment Inc.]*

Tidal Gravity and the *Endurance*'s Design

The *Endurance* has twelve modules linked in a ring, and a control module at the ring's center (Figure 20.1). Two landers and Rangers dock onto the *Endurance*'s central module.

In my scientist's interpretation of the movie, the *Endurance* was designed to survive strong tidal gravitational forces. This was important for the *Endurance*'s trip through the wormhole. The *Endurance* ring's diameter of 64 meters is nearly 1 percent of the wormhole's circumference. Steel and other solid materials break or flow, when subjected to distortions bigger than about a few tenths of a percent, so the dangers were obvious. And little was known about what the *Endurance* would encounter on the Gargantua side of the wormhole, so it was designed to withstand tidal forces far stronger than the wormhole's.

Now, a thin fiber can be bent around into complicated shapes without any portion of the fiber's material being distorted by anything close to 1 percent. The key is the thinness of the fiber. You could imagine the *Endurance*'s strength relying on a huge number of thin fibers stretching around the ring, like the strands of a cable that hold up a suspension bridge and can bend as needed when a strong wind blows. But that would make the ring *too* flexible. The ring needs much resistance to being deformed, so it won't deform so severely, when subjected to tidal forces, that the modules crash into each other.

The designers, in my interpretation, worked hard to make the *Endurance* resist deformation but be able to deform without breaking if it encounters tidal forces far stronger than anticipated.

Explosion in Orbit Above Mann's Planet

This design philosophy really pays off when Dr. Mann unwittingly triggers a huge explosion that breaks the *Endurance*'s ring, destroys two of the ring modules, and damages two others (Figure 20.2).

The explosion sets the ring spinning so fast that its modules feel 70 gees (70 Earth gravities) of centrifugal force. Its broken ends swing apart from each other but don't break, and the ring's modules don't crash into each other. This, in my scientist's interpretation, is a great example of conservative design by clever engineers!

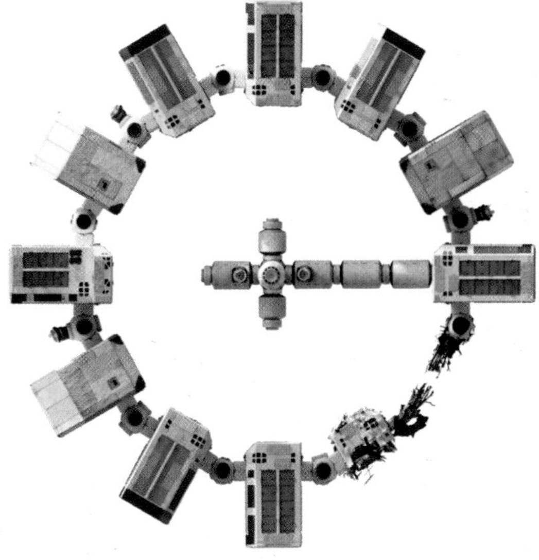

Fig. 20.2. *Left*: The explosion on the *Endurance*, with the lander above and Mann's planet below. [The ten radial light beams are lens flare due to scattering of light in the camera lens, not stuff from the explosion.] *Right*: The damaged *Endurance* after the explosion. *[From* Interstellar, *used courtesy of Warner Bros. Entertainment Inc.]*

Incidentally, I'm impressed by the explosion in the movie. An explosion in space makes no sound, as there is no air to transmit the sound waves. The *Endurance* explosion has no sound. The flames in such an explosion must quench quickly, as the oxygen that feeds them is quickly disbursed into space. The flames in the movie quench quickly. Paul Franklin tells me that his team worked hard to achieve this, as the explosion was a real one, on a movie set, and not a computer-generated visual effect. Another example of Christopher Nolan's commitment to science accuracy.

Our discussion of Gargantua's environs has taken us from the physics of planets (tidal deformation, tsunamis, tidal bores, . . .), through Gargantua's vibrations and the search for organic signs of life, to engineering issues (the *Endurance*'s robust design and its damaging explosion). As much as I enjoy these topics—and I've done research or textbook writing on most of them—they are not my greatest passion. My passion is extreme physics; physics at the edge of human knowledge and just beyond. That's where I take us next.

VI

EXTREME PHYSICS

21

The Fourth and Fifth Dimensions

Time as the Fourth Dimension

Ⓣ

In our universe, space has three dimensions: up-down, east-west, and north-south. But to schedule lunch with a friend, we must tell her not only where, but also *when*. In this sense, time is a fourth dimension.

However, time is a different kind of dimension than space. We have no trouble traveling westward as well as eastward; we make our choice and go. But having arrived at our luncheon, we cannot immediately, then and there, travel backward in time. No matter how hard we may try, we can only travel forward. The relativistic laws guarantee it. They enforce it.[1]

Nevertheless, time *is* a fourth dimension; it is *the* fourth dimension of our universe. The arena for our lives is four-dimensional spacetime, three space dimensions plus one time dimension.

When we physicists explore this spacetime arena by experiments and by mathematics, we discover that space and time are unified in several ways. At the simplest level, when we look out in space, we are

1 But the relativistic laws *do* offer the possibility of backward time travel by a circuitous route: going outward in space and returning before we left. To this I return in Chapter 30.

automatically looking backward in time because of how long it takes light to reach us. We see a quasar a billion light-years away as it was a billion years ago, when the light that enters our telescope was launched on its journey to us.

At a much deeper level, if you move relative to me at high speed, then we disagree on what events are simultaneous. You may think that two explosions, one on the Sun and the other on Earth are simultaneous, while I think the Earth explosion was five minutes before the one on the Sun. In this sense, what you regard as purely spatial (the separation of the explosions) I see as a mixture of space and time.

This mixing of space and time may seem counterintuitive, but it is fundamental to the very fabric of our universe. Fortunately, we can pretty much ignore it in this book except for Chapter 30.

The Bulk: Is It Real?

(EG)

Throughout this book, I visualize warped space by picturing our universe as a two-dimensional warped membrane, or brane, that resides in a bulk with three space dimensions—as in Figure 21.1, for example. Of course, in reality our brane has three space dimensions and the bulk has four, but I'm not very good at drawing that, so in my pictures I usually throw one dimension away.

Fig. 21.1. A small black hole spiraling into a large black hole, as viewed from the bulk with one space dimension removed. *[Drawing by Don Davis based on a sketch by me.]*

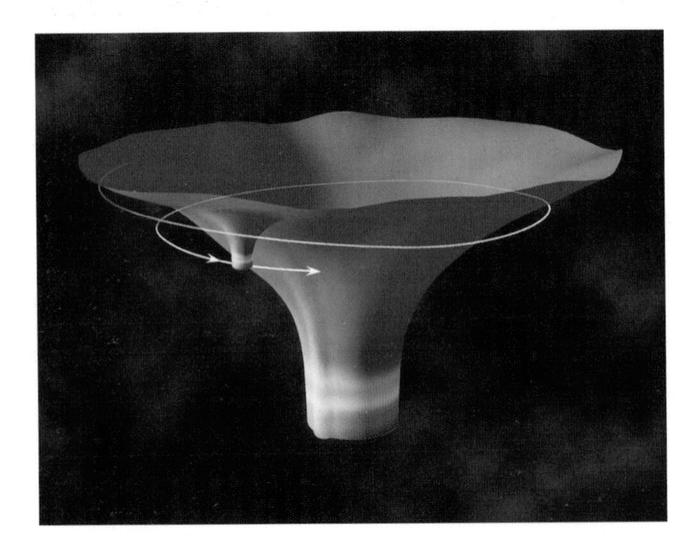

Does the bulk really and truly exist, or is it just a figment of our imaginations? Until the 1980s, most physicists, including me, thought it a figment.

How could it be a figment? Don't we know for sure that our universe's space is warped? Don't the radio signals sent to the *Viking* spacecraft reveal its warpage to high precision (Chapter 4)? *Yes. . . .* And since our space is truly warped, doesn't it have to be warped inside some higher-dimensional space, inside some bulk?

No. It is perfectly possible for our universe to be warped without there really existing a higher-dimensional bulk. We physicists can describe our universe's warping, in mathematics, without the aid of a bulk. We can formulate Einstein's relativistic laws, which govern the warping, without the aid of a bulk. In fact, that's how we almost always do it, in our research. The bulk, for us, until the 1980s, was just a visual aid. An aid to give us intuition about what's going on in our mathematics, and to help us communicate with each other and with nonphysicists. A visual aid. Not a real thing.

What would it mean for the bulk to be real? How can we test whether it's real? The bulk is real if it can influence things we measure. And until the 1980s we saw no way it could influence our measurements.

Then in 1984 this changed. Radically. Michael Green at the University of London and John Schwarz at Caltech had a huge breakthrough in the quest to discover the laws of quantum gravity.[2] But strangely, their breakthrough worked only if our universe is a brane embedded in a bulk that has one time dimension and nine space dimensions—a bulk with six more space dimensions than our brane. In the mathematical formalism that Green and Schwarz were pursuing, called "superstring theory," the bulk's extra dimensions influence our brane in major ways, in ways that can be measured in physics experiments when we have sufficiently advanced technology. In ways that may make it possible to reconcile the laws of quantum physics with Einstein's relativistic laws.

Fig 21.2. *Left*: Michael Green (*left*) and John Schwarz hiking in Aspen, Colorado, in 1984, at the time of their breakthrough. *Right*: Michael Green (*left*) and John Schwarz (*right*) being awarded the three-million-dollar 2014 Fundamental Physics Prize for their breakthrough. In the middle are Yuri Milner (founder of the prize) and Mark Zuckerberg (Facebook cofounder).

2 See Chapter 3 for a brief description of this quest.

Since the Green-Schwarz breakthrough, we physicists have taken superstring theory very seriously and have put great effort into exploring and extending it. And, consequently, we have taken very seriously the idea that the bulk truly exists and truly can influence our universe.

The Fifth Dimension

(EG)

Although superstring theory says the bulk has six more dimensions than our universe, there is reason to suspect that, for practical purposes, the number of extra dimensions is really only one. (I explain this in Chapter 23.)

For this reason, and because six extra dimensions is a bit much for a science-fiction movie, *Interstellar*'s bulk has just one extra dimension, for a total of five dimensions in all. It shares three space dimensions with our brane: east-west, north-south, and up-down. It shares a fourth, time dimension, with our brane. And it has a fifth space dimension, *out-back*, which extends perpendicular to our brane, both above the brane and below, as depicted in Figure 21.3.

The out-back dimension plays a major role in *Interstellar*, though the Professor and others don't use the phrase "out-back," but instead just refer to "the fifth dimension." Out-back is central to the next two chapters, and to Chapters 25, 29, and 30.

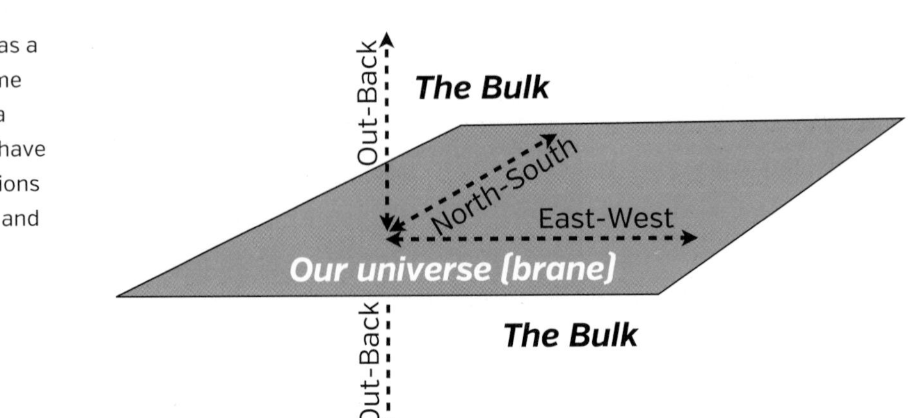

Fig. 21.3. Our universe as a brane with four spacetime dimensions, residing in a five-dimensional bulk. I have suppressed two dimensions from the diagram: time, and our universe's up-down dimension.

22

Bulk Beings

2D Brane and 3D Bulk

Ⓣ

In 1844 Edwin Abbott wrote a satirical novella titled *Flatland: A Romance of Many Dimensions* (Figure 22.1).[1] Though its satire on Victorian culture seems quaint today and its attitude toward women outrageous, the novella's venue is highly relevant to *Interstellar.* I recommend it to you.

It describes the adventures of a square-shaped being who lives in a two-dimensional universe called Flatland. The square visits a one-dimensional universe called Lineland, a zero-dimensional universe called Pointland, and most amazing of all to him, a three-dimensional universe called Spaceland. And, while living in Flatland, he is visited by a spherical being from Spaceland.

In my first meeting with Christopher Nolan, we

Fig. 22.1. The cover of the first edition of *Flatland*.

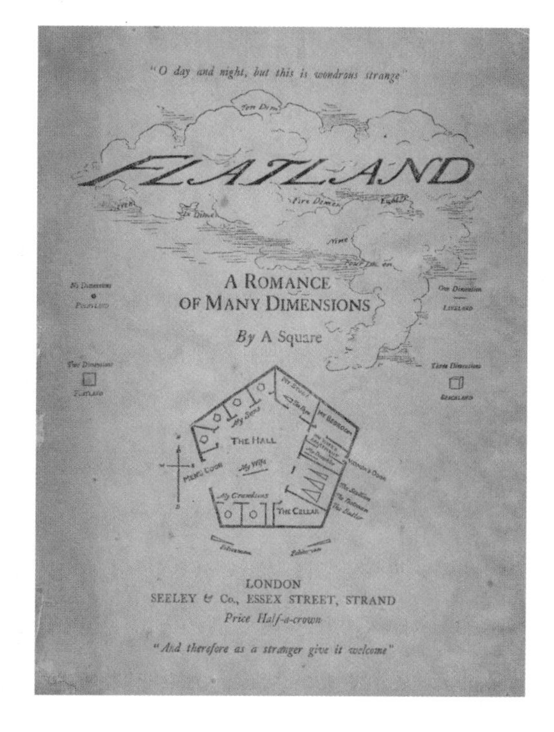

1 Widely available on the web. See, for example, the end of the article "Flatland" on Wikipedia.

were both delighted to find the other had read Abbott's novella and loved it.

In the spirit of Abbott's novella, imagine that you are a two-dimensional being, like the square, who lives in a two-dimensional universe like Flatland. Your universe could be a tabletop, or a flat sheet of paper, or a rubber membrane. In the spirit of modern physics, I refer to it as a *two-dimensional (2D) brane.*

Being well educated, you suppose there is a 3D bulk, in which your brane is embedded, but you're not certain. Imagine your excitement when one day you are visited by a sphere from the 3D bulk. A "bulk being," you might call him.

At first you don't realize it's a bulk being, but after much observation and thought, you see no other explanation. What you observe is this: Suddenly, with no warning and no apparent source, a blue point appears in your brane (top left of Figure 22.2). It expands to become a blue circle that grows to a maximum diameter (middle left), and then gradually shrinks to a point (bottom left) and disappears completely.

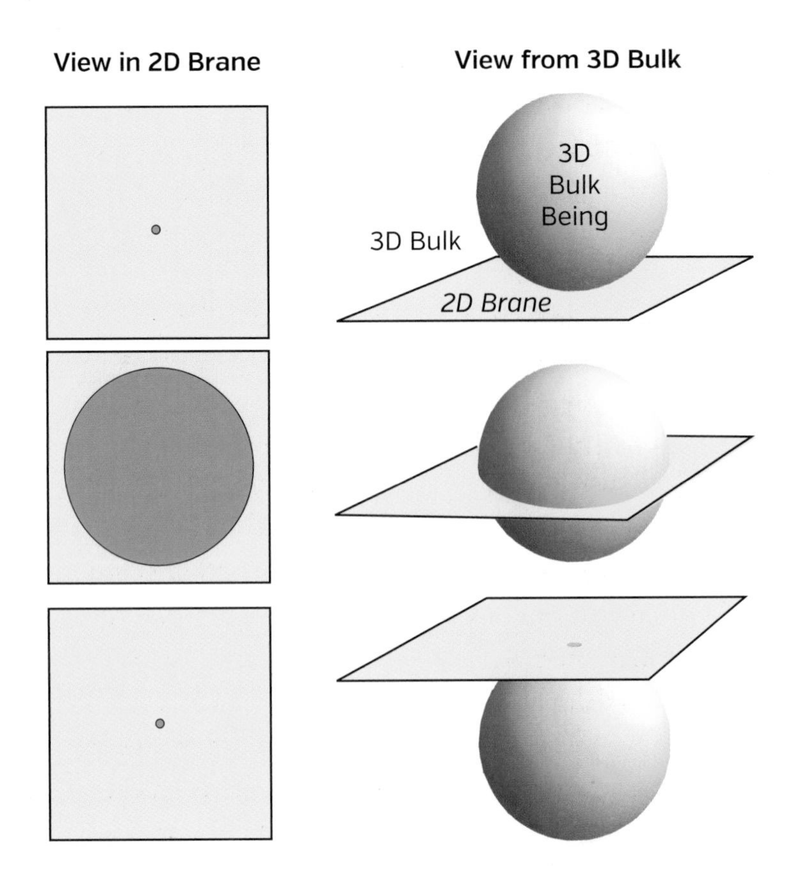

Fig. 22.2. A three-dimensional sphere passes through a two-dimensional brane.

You believe in conservation of matter. No object can ever be created from nothing, yet this object was. The only explanation you can find is shown in the right half of Figure 22.2. A three-dimensional bulk being—a sphere—passed through your brane. As it passed through, you saw in your brane its changing two-dimensional cross section. The cross section began with a point at the sphere's south pole (top right). It expanded to a maximal circle, the sphere's equatorial plane (middle right). It then shrank to a point, the sphere's north pole, and disappeared (bottom right).

Imagine what would happen if a 3D human being, living in the 3D bulk, passed through your 2D brane. What would you see?

Bulk Beings from the Fifth Dimension, Passing Through Our 3D Brane

Ⓣ

Suppose that our universe, with its three space and one time dimensions, really does live in a five-dimensional bulk (four space and one time). And suppose there are "hyperspherical beings" who live in the bulk. Such a being would have a center and a surface. Its surface would consist of all points, in four space dimensions, that are some fixed distance from the center, for example, 30 centimeters. The bulk being's surface would have three dimensions and its interior would have four.

Suppose that this hyperspherical bulk being, traveling in the bulk's out direction or back direction, were to pass through our brane. What would we see? The obvious guess is correct. We would see spherical cross sections of the hypersphere (Figure 22.3).

A point would appear from nothing (1). It would expand to become a three-dimensional sphere (2). The sphere would expand to a maximum diameter (3), then contract (4), shrink to a point (5), and disappear.

Can you guess what we would see if a four-dimensional human being living in the bulk were to pass through our brane? To speculate about this, you need to imagine what a

Fig. 22.3. A hyperspherical bulk being passing through our brane, as seen in our brane.

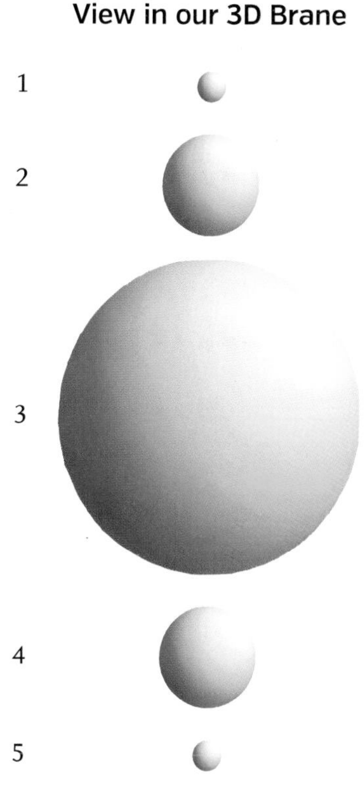

four-dimensional human being—with two legs, a torso, two arms, and a head—must "look like" in the bulk, with its four space dimensions. And what its cross sections must look like.

The Nature of Bulk Beings, and Their Gravity

(EG) & (S)

If there *are* bulk beings, what are they made of? Certainly not atom-based matter like us. Atoms have three space dimensions. They can only exist in three space dimensions, not four. And this is true of subatomic particles as well. And it is true also of electric fields and magnetic fields (Chapter 2) and the forces that hold atomic nuclei together.

Some of the world's most brilliant physicists have struggled to understand how matter and fields and forces behave if our universe really is a brane in a higher-dimensional bulk. Those struggles have pointed rather firmly to the conclusion that all the particles and all the forces and all the fields known to humans are confined to our brane, with one exception: gravity, and the warping of spacetime associated with gravity.

There might be other kinds of matter and fields and forces that have four space dimensions and reside in the bulk. But if there are, we are ignorant of their nature. We can speculate. Physicists do speculate. But we have no observational or experimental evidence to guide our speculations. In *Interstellar,* on Professor Brand's blackboard, we see him speculating (Chapter 25).

It's a reasonable, half-educated guess that, if bulk forces and fields and particles do exist, we will never be able to feel them or see them. When a bulk being passes through our brane, we will not see the stuff of which the being is made. The being's cross sections will be transparent.

On the other hand, we *will* feel and see the being's gravity and its warping of space and time. For example, if a hyperspherical bulk being appears in my stomach and has a strong enough gravitational pull, my stomach may begin to cramp as my muscles tighten, trying to resist getting sucked to the center of the being's spherical cross section.

If the bulk being's cross section appears and then disappears in front of a checkerboard of paint swatches, its space warp might lens the swatches, bending the image I see, as in the top half of Figure 22.4.

And if the bulk being is spinning, it might drag space into a whirling motion that I can feel and see, as in the bottom of Figure 22.4.

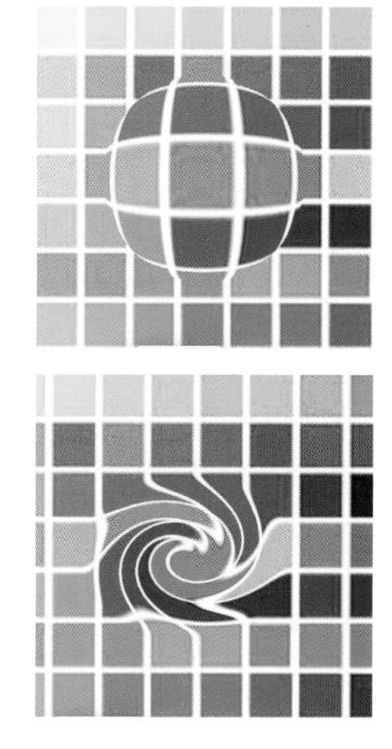

Fig. 22.4. A bulk being, passing through our brane, bends and swirls our view of a paint-swatch wall.

Interstellar's Bulk Beings

All the characters in *Interstellar* are convinced that bulk beings exist, though they use that name only rarely. Usually, the characters call the bulk beings "They." A reverential They. Early in the movie, Amelia Brand says to Cooper, "And whoever They are, They appear to be looking out for us. That wormhole lets us travel to other stars. It came along right as we needed it."

One of Christopher Nolan's clever and intriguing ideas is to imagine that They are actually our descendants: humans who, in the far future, evolve to acquire an additional space dimension and live in the bulk. Late in the movie, Cooper says to TARS, "Don't you get it yet, TARS? They aren't beings. They're us, trying to help, just like I tried to help Murph." TARS responds, "People didn't build this tesseract" (in which Cooper is riding; Chapter 29). "Not yet," Cooper says, "but one day. Not you and me but people, people who've evolved beyond the four dimensions we know."

Cooper, Brand, and the crew of the *Endurance* never actually feel or see our bulk descendants' gravity or their space warps and whirls. That, if it ever occurs, is left for a sequel to *Interstellar*. But older Cooper himself, riding through the bulk in the closing tesseract of Chapter 30, reaches out to the *Endurance*'s crew and his younger self, reaches out through the bulk, reaches out gravitationally. Brand feels and sees his presence, and thinks he is They.

23

Confining Gravity

The Trouble with Gravity in Five Dimensions

(EG)

If the bulk *does* exist, then its space *must* be warped. If it were not warped, then gravity would obey an inverse cube law instead of inverse square, our Sun could not hang onto its planets, and the solar system would fly apart.

Fig. 23.1. The gravitational force lines around the Sun.

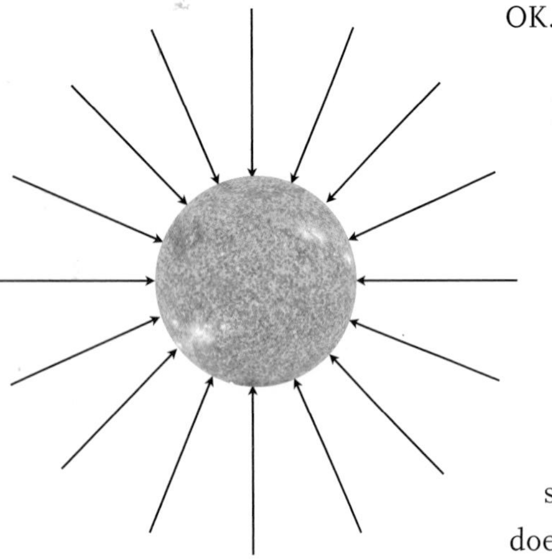

OK. I'll slow down and explain this more carefully.

Recall (Chapter 2) that the Sun's gravitational force lines, like those of the Earth and any other spherical body, point radially toward its center and pull objects along themselves toward the Sun (Figure 23.1). The strength of the Sun's gravitational pull is proportional to the density of the force lines (the number of lines passing through a fixed area). And since the transverse areas (spheres) through which the lines pass have two dimensions, the lines' density goes down with increasing radius r as $1/r^2$, and so does gravity's strength. This is *Newton's inverse square law* for gravity.

String theory insists that gravity in the bulk is also described by

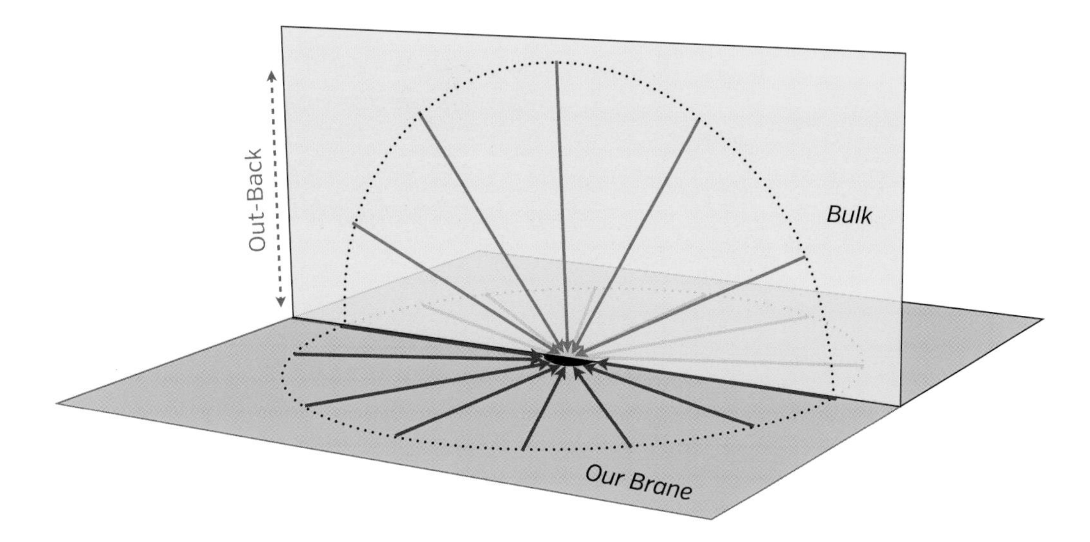

Fig. 23.2. Gravitational force lines spread radially into the bulk, if the bulk is not warped. The dotted circles are solely to guide your eyes. *[Patterned on a figure in Lisa Randall's* Warped Passages *(Randall 2006).]*

force lines. If space in the bulk is not warped, then the Sun's gravitational force lines will spread radially outward into the bulk (Figure 23.2). Because of the bulk's extra dimension (just one in *Interstellar*), there are three transverse dimensions into which gravity can spread instead of just two. Therefore, *if the bulk exists and is not warped, then the density of force lines and thence gravity's strength should decrease as* $1/r^3$ *when we move away from the Sun, rather than as* $1/r^2$. The sun's pull on the Earth would be two hundred times weaker, and on Saturn 2000 times weaker. With gravity weakening so rapidly, the Sun couldn't hold onto its planets; they would fly away into interstellar space.

But they don't fly away. And their measured motions reveal unequivocally that the Sun's gravity weakens as the inverse square of the distance. The conclusion is inescapable: if there *is* a bulk, it must be warped in some manner that prevents gravity from spreading into the fifth dimension, the *out-back* dimension.

Is Out-Back Curled Up?

(EG)

If the bulk's out-back dimension were *curled up* into a tight roll, then gravity could not spread far into the bulk, and the inverse square law would be restored.

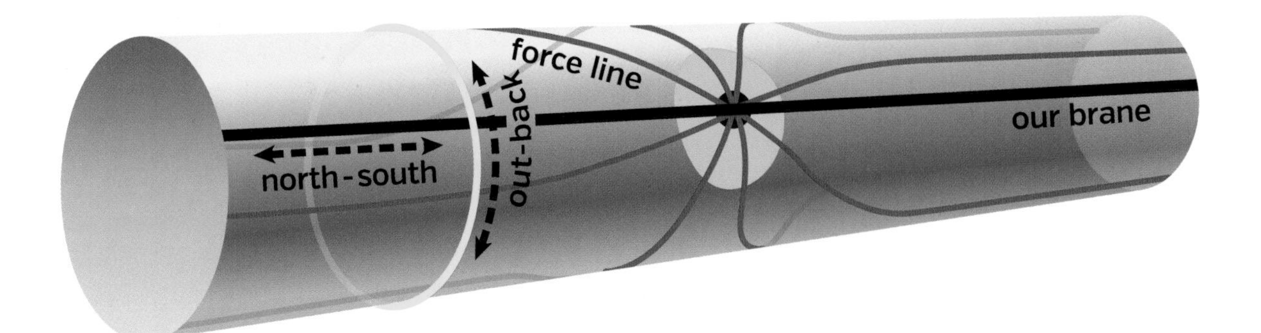

Fig. 23.3. If the out-back dimension (*yellow*) is curled up, then outside the blue circle a particle's gravitational force lines (*red*) are parallel to our brane.

Figure 23.3 depicts this for the gravity of a tiny particle that resides at the center of the blue disk. In this picture, two space dimensions are suppressed, so we see only one of our brane's dimensions (call it north-south) along with the bulk's out-back dimension. Near the particle, inside the blue disk, the force lines spread in the out-back dimension as well as north-south, so (with the missing dimensions restored) gravity's strength obeys an inverse cube law. However, outside the blue disk the curl-up makes the force lines lie parallel to our brane. They spread no further into out-back, and Newton's inverse square law is restored.

Physicists who struggle to understand quantum gravity think this is the fate of all the extra dimensions except possibly one or two: they are curled up on microscopic scales, preventing gravity from spreading too fast. In *Interstellar,* Christopher Nolan ignores these curled-up dimensions and focuses on just one bulk dimension that's not curled up. This becomes his out-back, fifth dimension.

Why should out-back not be curled up? For Chris the answer is simple: A curled-up bulk has very little volume—nowhere near enough volume to be an arena for interesting science fiction. For Cooper to travel into the bulk riding in the tesseract, as he does in the movie, the tesseract needs far more volume than a curled-up dimension would provide.

Out-Back: The Anti-DeSitter Warp

(EG)

In 1999, Lisa Randall at Princeton University and MIT and Raman Sundrum at Boston University (Figure 23.4) conceived another way to

Fig. 23.4. Lisa Randall (1962– , right) and Raman Sundrum (1964– , left).

stop gravitational force lines from spreading into the bulk: the bulk could suffer what is called "Anti-deSitter warping." This warping might be produced by what are called "quantum fluctuations of bulk fields"—but that's irrelevant to my story so I do not explain it here.[1] Suffice it to say that this mechanism to produce the warping is very natural. By contrast, the Anti-deSitter (AdS) warping itself does not *look* natural at all. It looks downright weird.

Suppose you're a microbe, and you live in a face of a microscopic tesseract (Chapter 29). You travel, in your tesseract, out from our brane; perpendicularly out (straight up in Figure 23.5). And suppose you have a microbial pal, who also travels perpendicularly out from our brane. When you and your pal depart our brane, you are 1 kilo-

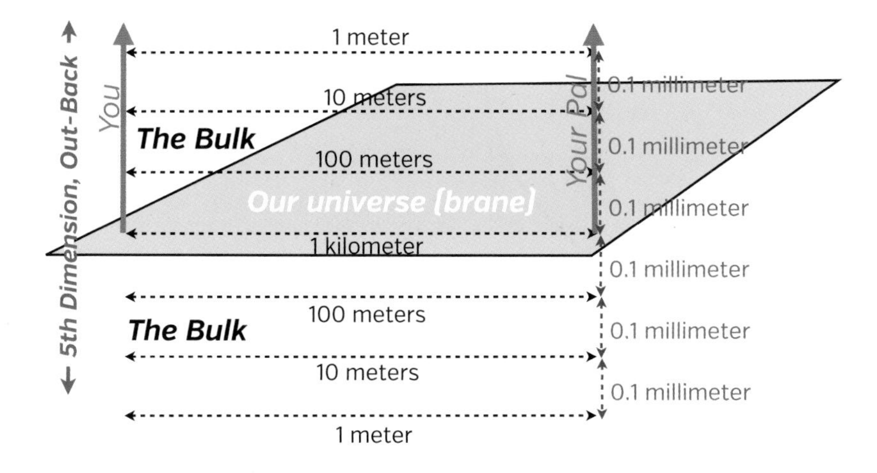

Fig. 23.5. AdS warping of the bulk.

1 I discuss quantum fluctuations in Chapter 26 and bulk fields in Chapter 25.

meter apart (1000 meters; about 0.6 miles). Although you both travel precisely outward, perpendicular to our brane, your separation plummets precipitously due to AdS warping. When you have traveled a tenth of a millimeter (the thickness of a human hair), your separation has decreased tenfold: from 1 kilometer to 100 meters. The next 0.1 millimeter of travel reduces your separation by another factor of ten, to 10 meters; the next 0.1 millimeter reduces it to 1 meter; and so forth.

This shrinkage of distances parallel to our brane is hard to imagine. I don't know a good way to draw it, no better way than Figure 23.5. But it has marvelous consequences.

It has the potential to explain a mystery called the "hierarchy problem in the laws of physics"—but that's outside the scope of this book.[2] And because of the shrinkage, there is very little volume, above or below our brane, into which gravitational force lines can spread (Figure 23.6). Closer to our brane than 0.1 millimeter, the force lines spread into three transverse dimensions with impunity, so gravity obeys an inverse cube law. Farther than 0.1 millimeter, the force lines are bent parallel to our brane and so spread into just two transverse dimensions, whence gravity obeys the observed inverse square law.[3]

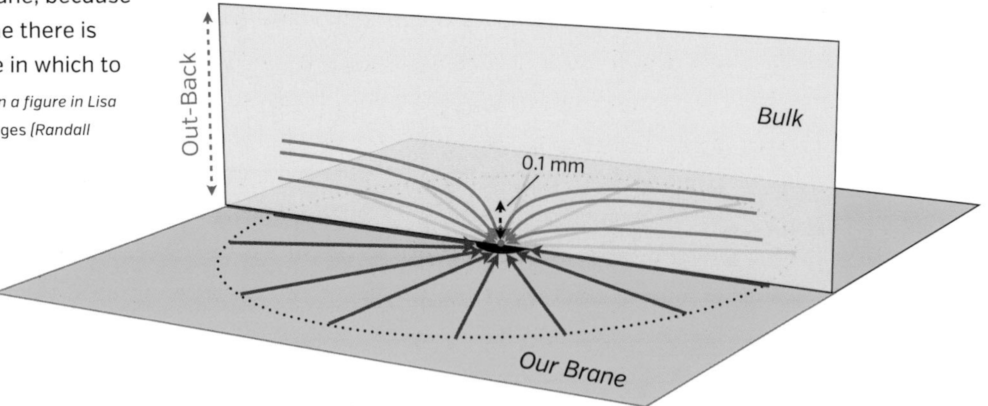

Fig. 23.6. If the bulk experiences AdS warping, then gravitational force lines bend parallel to our brane, because far from the brane there is very little volume in which to spread. *[Patterned on a figure in Lisa Randall's* Warped Passages *[Randall 2006].]*

2 For details see Lisa Randall's *Warped Passages* (HarperCollins, 2006).

3 Why is the magic distance, at which the inverse square law begins, 0.1 millimeter instead of, say, 1 kilometer or 1 picometer? I have chosen 0.1 millimeter quite arbitrarily. Experiments have proved that gravity obeys the inverse square law down to about 0.1 millimeter, so that is an upper limit on the magic distance. It could perfectly well be smaller.

The AdS Sandwich: Plenty of Room in the Bulk

Ⓢ

Sadly, the precipitous shrinkage of distances parallel to our brane, as you move outward, makes the bulk's volume above and below our brane too small for Cooper and his tesseract, and too small for any other human activity in the bulk. I recognized this problem way back in 2006, when *Interstellar* was in its infancy, and I quickly conceived a solution for my science interpretation of the movie: Confine the AdS warping to a thin layer around our brane, a "sandwich." Do so by placing two other branes, confining branes, alongside ours (Figure 23.7). In the sandwich between these branes, the bulk suffers AdS warping. Outside the sandwich, the bulk is totally unwarped. So there is all the volume any sci-fi writer could want, outside the sandwich, for bulk-based adventures.

How thick must the sandwich be? Thick enough to bend gravitational force lines—emerging from our brane—parallel to our brane and hold them there, so we in our brane see gravity obey an inverse square law. But no thicker, because added thickness means greater total transverse shrinkage, which may cause trouble for bulk-based adventures. (Suppose our whole universe, as seen from outside the AdS layer, were shrunk to the size of a pin head!) The required thickness turns out to be about 3 centimeters (roughly an inch), so as you travel from our brane to a confining brane, distances parallel to our brane shrink by fifteen powers of ten: a thousand trillion.

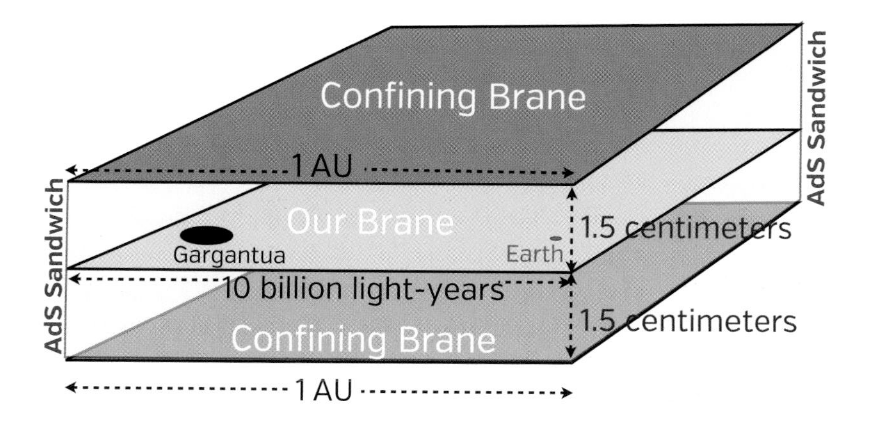

Fig. 23.7. The AdS sandwich between two confining branes. The AdS layer between the branes is lightly grayed.

In my interpretation of *Interstellar*, Gargantua is in the far reaches of the observable universe: roughly 10 billion light-years from Earth. Cooper, in the tesseract, rises through the AdS layer, from Gargantua's core into the bulk. There the distance to Earth is 10 billion light-years divided by a thousand trillion, which is about the same as the distance between the Sun and the Earth, one "astronomical unit" (1 AU; Figure 23.7). Cooper then travels that 1 AU distance through the bulk, parallel to our brane, to reach the Earth and visit Murph; see Figure 29.4.

DANGER: The Sandwich Is Unstable

Ⓢ

In 2006, I used Einstein's relativistic laws to work out a mathematical description of the AdS layer and its confining branes. Because I had never before worked with relativity in five dimensions, I asked Lisa Randall to critique my analysis. Lisa browsed it quickly, and then told me some good news and some bad news.

The good news: My idea of an AdS sandwich had been invented six years earlier by Ruth Gregory (University of Durham, UK), together with Valery Rubakov and Sergei Sibiryakov (Institute for Nuclear Research in Moscow, Russia). This showed I was not being stupid in my first mathematical foray into the bulk. I had rediscovered something worth discovering.

The bad news: Edward Witten (Princeton) and others had shown that the AdS sandwich is *unstable!* The confining branes are under pressure, rather like a playing card that you squeeze end to end between your finger and thumb (Figure 23.8). The card bends, and with further squeezing, it buckles. Similarly, the confining branes will bend and crash into our brane (our universe), destroying it. The entire universe destroyed! That's the worst news ever!!

But I can think of several ways to save our universe, if it really does live in an AdS sandwich (which I very much doubt it does); several ways to "stabilize the confining branes," in the jargon of physicists.

In my science interpretation of *Interstellar*, Professor Brand, working with Einstein's relativity equations, rediscovers the AdS sandwich, as I

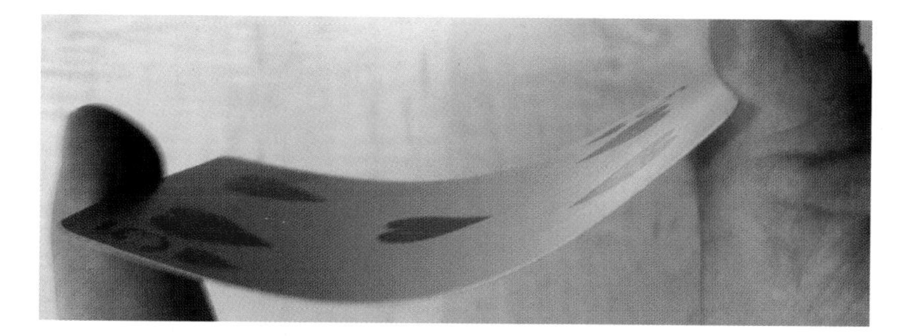

Fig. 23.8. A playing card, compressed end to end, bends and then buckles.

did; see the photograph of his blackboard in Figure 3.6. How the confining branes are stabilized then gets intertwined with the Professor's struggle to understand and control gravitational anomalies. In the movie, that struggle is spelled out mathematically on the sixteen blackboards in Professor Brand's office; Chapter 25.

Traveling Through the AdS Layer

Ⓢ

In the AdS layer, the AdS warpage of space produces tidal forces that are enormous by human standards. Any bulk being traveling through the layer to reach our brane must deal with those forces. Because we know nothing about the matter of which a bulk being is made—matter with four space dimensions—we have no idea whether this is an issue. In science fiction it can be left in the hands of the writers.

Not so for Cooper, riding in the tesseract (Chapter 29). In my interpretation of the movie, he has to cross the AdS layer. The tesseract must either protect him from the layer's enormous tidal forces or clear the AdS layer away from his path. Otherwise he'll be spaghettified.

By confining gravity, the AdS layer regulates its strength. In *Interstellar* we see gravity's strength fluctuate, perhaps due to fluctuations in the AdS layer. These fluctuations—gravitational anomalies—play a huge role in *Interstellar*. To them we now turn.

24

Gravitational Anomalies

A *gravitational anomaly* is something about gravity that doesn't fit our understanding of the universe, or our understanding of the physical laws that control the universe—for example the falling books, in *Interstellar,* that Murph attributes to a ghost.

Since 1850, physicists have put a lot of effort into searching for gravitational anomalies and understanding those few that were found. Why? Because any true anomaly is likely to produce a scientific revolution; a major change in what we think is True Ⓣ. This, in fact, has happened three times since 1850.

In *Interstellar,* Professor Brand's struggle to understand gravitational anomalies is very much in the spirit of these previous revolutions; so I describe the previous ones, briefly.

The Anomalous Precession of Mercury's Orbit

Ⓣ

Newton's inverse square law for gravity (Chapters 2 and 23) forces the orbits of the planets around the Sun to be ellipses. Each planet feels small gravitational tugs from the other planets, and these tugs cause its ellipse to gradually change orientation, that is, to gradually *precess.*

In 1859, the astronomer Urbain Le Verrier at the Observatoire de Paris (France) announced he had discovered an anomaly in the orbit of the planet Mercury. When he computed the total precession of Mercury's orbit caused by all the other planets, he got the wrong answer. The measured precession is larger than the planets could produce by about 0.1 arc second each time Mercury traverses its orbit (Figure 24.1).

Now 0.1 arc second is a tiny angle, just one ten-millionth of a circle. But Newton's inverse square law insists there can be no anomaly whatsoever.

Le Verrier convinced himself that this anomaly is produced by the gravitational tug of an undiscovered planet closer to the Sun than Mercury; "Vulcan" he called it.

Astronomers searched in vain for Vulcan. They could not find it, nor could they find any other explanation for the anomaly. By 1890 the conclusion seemed clear: Newton's inverse square law must be very slightly wrong.

Wrong in what way? A revolutionary way, it turned out. The way discovered by Einstein twenty-five years later. The warping of time and space endow the Sun with a gravitational force that obeys Newton's inverse square law, but only nearly. Not precisely.

Upon realizing that his new relativistic laws explain the observed anomaly, Einstein was so excited that he suffered heart palpitations and felt like something snapped inside himself. "For a few days I was beside myself with joyous excitement."

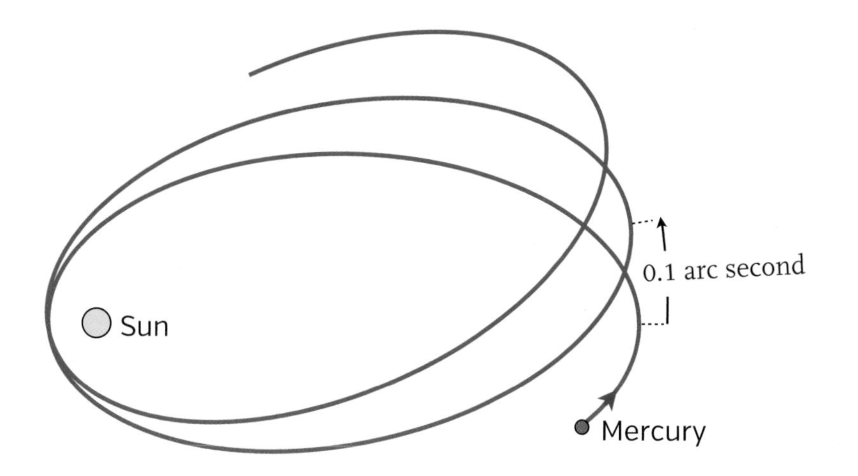

Fig. 24.1. The anomalous precession of Mercury's orbit. In this picture, I exaggerate the orbit's ellipticity (its elongated shape) and the magnitude of its precession.

Today the measured anomalous precession and the predictions by Einstein's laws agree to within one part in a thousand (one-thousandth of the anomalous precession), which is the accuracy of the observations—a great triumph for Einstein!

The Anomalous Orbits of Galaxies Around Each Other

Ⓣ

In 1933 the Caltech astrophysicist Fritz Zwicky announced he had discovered a huge anomaly in the orbits of galaxies around each other. The galaxies were in the Coma cluster (Figure 24.2), a collection of about a thousand galaxies, 300 million light-years from Earth, in the constellation Coma Berenices.

From the Doppler shifts of the galaxies' spectral lines, Zwicky could estimate how fast they were moving relative to each other. And from the brightness of each galaxy, he could estimate its mass and thence its

Fig. 24.2. The Coma cluster of galaxies as seen through a large telescope.

gravitational pull on the other galaxies. The galaxies' motions were so fast that there was no way their gravitational pulls could hold the cluster together. Our best understanding of the universe and of gravity insisted that the cluster must be flying apart, and would soon be completely destroyed. If so, then the cluster must have formed by random motions of all those galaxies and would disrupt in a veritable blink of an eye compared to other astronomical phenomena.

This conclusion was totally implausible to Zwicky. Something was wrong with our conventional wisdom. Zwicky made an educated guess: The Coma cluster must be filled with some sort of "dark matter" whose gravity is strong enough to hold the cluster together.

Now, many anomalies that astronomers and physicists think they have discovered go away when observations improve. This one did not. Instead, it spread. By the 1970s it was clear that so-called dark matter permeates most all clusters of galaxies and even individual galaxies. By the 2000s, it was clear that the dark matter gravitationally lenses light from more distant galaxies (Figure 24.3), just as Gargantua gravitationally lenses light from stars (Chapter 8). Today that lensing is being used to map the dark matter in our universe.

Fig. 24.3. Dark matter in the galaxy cluster Abell 2218 gravitationally lenses more distant galaxies. The images of the lensed galaxies are arc-shaped (e.g., those I circled in purple), analogous to arc structures seen in Gargantua's gravitational lensing, Chapter 8.

And today physicists are fairly sure that the dark matter is truly revolutionary, that it consists of fundamental particles of a type never before seen, but a type predicted by our best current understanding of the quantum laws of physics. Physicists have embarked on a holy-grail mission: a quest to detect these particles of dark matter, shooting through the Earth with near impunity, and measure their properties.

The Anomalous Acceleration of the Universe's Expansion

Ⓣ

In 1998 two research groups independently discovered an astounding anomaly in the expansion of our universe. For this discovery, the groups' leaders (Saul Perlmutter and Adam Reiss at the University of California, Berkeley, and Brian Schmidt at the Australian National University) won the 2011 Nobel Prize in Physics.

Fig. 24.4. The distance to the star at the time of explosion (the time that the light we see was emitted), under two assumptions: that the universe's expansion is decelerating (*red*) or accelerating (*blue*). The explosion was dimmer than expected, so farther away. The universe must be accelerating.

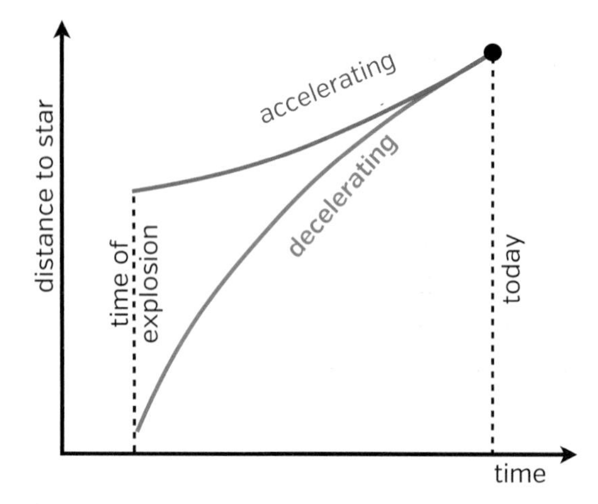

Both groups were observing supernova explosions: explosions triggered when a massive star exhausts its nuclear fuel and implodes to form a neutron star, and the implosion energy blows off the star's outer layers. They discovered that distant supernovae are dimmer than expected, and therefore farther away than expected. Farther enough away that the universe's expansion must have been slower in the past than today. The expansion is accelerating. See Figure 24.4.

Now, our best understanding of gravity and the universe required, unequivocally, that all things in the universe (stars, galaxies, galaxy clusters, dark matter, etc.) must *pull* on each other gravitationally. And by that pull they must *slow* the universe's expansion. The universe's expansion must slow down over time, not speed up.

For this reason, I, personally, didn't believe the claimed acceleration, nor did many of my astronomer and physicist colleagues. We didn't believe until other observations, by completely different methods, confirmed it. Then we caved.

So what's going on? There are two possibilities: Something is wrong with Einstein's relativistic laws of gravity. Or something else is filling the universe, in addition to ordinary matter and dark matter. Something that *repels* gravitationally.

Most physicists love Einstein's relativistic laws and are loathe to give them up, and so lean toward repulsion. The hypothetical material that repels has been given the name "dark energy."

The final verdict is not in. But if the cause of the anomaly is, indeed, dark energy (whatever that may be), then gravitational observations now tell us that 68 percent of the universe's mass is in dark energy, 27 percent is in dark matter, and only 5 percent is in the kind of ordinary matter of which you, I, planets, stars, and galaxies are made.

So physicists today have another holy grail: to understand whether the universe's accelerated expansion is caused by a breakdown of Einstein's relativistic laws (and if so, what is the nature of the correct laws?), or is caused by repulsive dark energy (and if so, what is the nature of the dark energy?).

Gravitational Anomalies in *Interstellar*

The gravitational anomalies in *Interstellar* are seen on Earth, by contrast with the three anomalies that I described.

Physicists have put great effort into searching for such anomalies on Earth, beginning with Isaac Newton himself in the late 1600s. Those searches have produced many claimed anomalies, but all claims, upon deeper scrutiny, have collapsed.

The anomalies in *Interstellar* are startling for their weirdness and strength, and the way they change as time passes. If anything like them had occurred in the twentieth century or early twenty-first, physicists would surely have noticed them and explored them with great fervor. Somehow, gravity on Earth has been altered in the era of *Interstellar.*

And, indeed, Romilly tells Cooper so in the movie: "We started detecting gravitational anomalies [on Earth] almost fifty years ago," and also, around that same time, the most significant anomaly of all: the sudden appearance of a wormhole near Saturn, where before there was none.

In the movie's opening scene, Cooper experiences an anomaly himself, while trying to land a Ranger spacecraft. "Over the Straights something tripped my fly-by-wire," he tells Romilly.

The GPS system that Cooper has adapted to control harvesting machines, as they roam through corn fields, has also gone haywire, and a bunch of harvesters have converged on his farmhouse. He attributes this to gravitational anomalies that screwed up the gravity corrections that any GPS system relies on (Figure 4.2).

Early in the movie, we see Murph watch, transfixed, as dust falls unnaturally fast to the floor of her bedroom, collecting in a bar-code-like pattern of thick lines. And then we see Cooper stare at the lines (Figure 24.5) and toss a coin across one. The coin shoots to the floor.

In my science interpretation of *Interstellar*, I presume that Profes-

Fig. 24.5. Cooper stares at the dust pattern on the floor of Murph's bedroom. *[From* Interstellar, *used courtesy of Warner Bros. Entertainment Inc.]*

sor Brand's team has collected a large trove of data on the anomalies. The most interesting data to me as a physicist, and to Professor Brand in my movie interpretation, is new and changing patterns of *tidal gravity*.

We first met tidal gravity in Chapter 4: the tidal gravity produced by a black hole, and tidal gravity on Earth produced by the Moon and Sun. In Chapter 17 we saw Gargantua's tidal gravity in action on Miller's planet, triggering gigantic "Millerquakes," tsunamis, and tidal bores. In Chapter 16 we met the tiny stretching and squeezing of tidal gravity in a gravitational wave.

Tidal gravity is produced not only by black holes, the Sun, the Moon, and gravitational waves but also, in fact, by all gravitating objects. For example, regions of the Earth's crust that contain oil are less dense than regions containing only rock, so their gravitational pull is weaker. This leads to a peculiar pattern of tidal gravitational forces.

In Figure 24.6, I use tendex lines to illustrate that tidal-force pattern. (See Chapter 4 for a discussion of tendex lines.) Squeezing tendex lines (drawn blue) stick out of the oil-bearing region, while stretching tendex lines (drawn red) stick out of the denser, oil-free region. As always, the two families of tendex lines are perpendicular to each other.

An instrument called a gravity gradiometer can measure these tidal

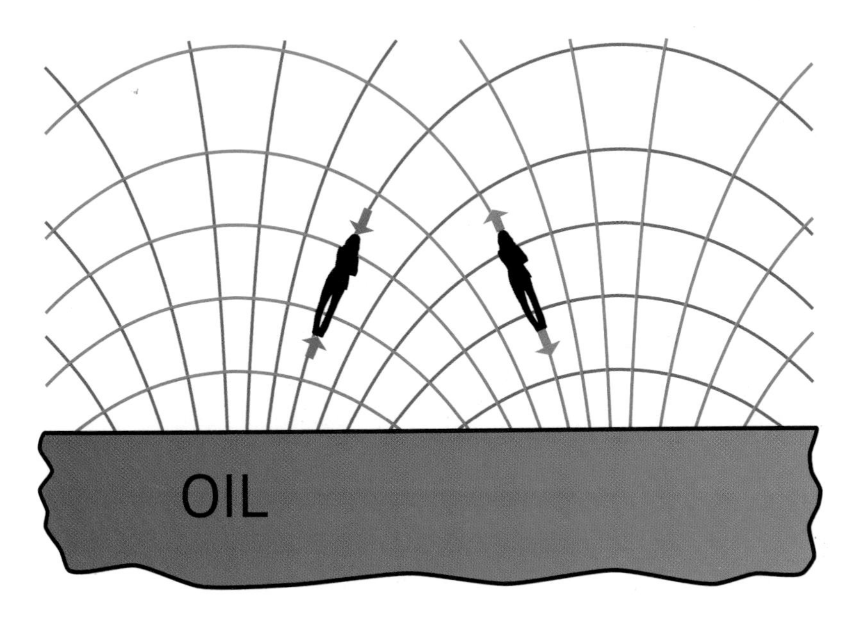

Fig. 24.6. Tendex lines above a portion of the Earth's crust. The red lines produce a tidal stretch along themselves. The blue lines produce a tidal squeeze.

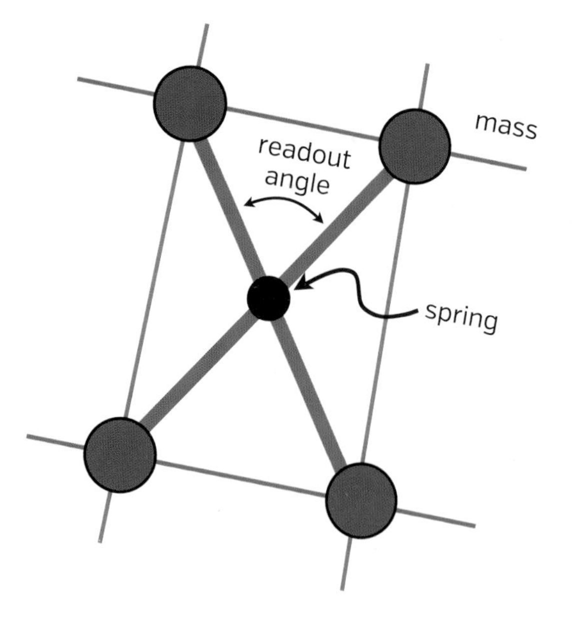

Fig. 24.7. A simple version of a gravity gradiometer, designed and built by Robert Forward at Hughes Research Laboratories in 1970.

patterns (Figure 24.7). It consists of two crossed, solid rods attached to a torsional spring. On the ends of each rod are masses that feel gravity. The rods are normally perpendicular to each other, but in the figure the blue tendex lines squeeze the top two masses together and squeeze the bottom two together, while the red tendex lines stretch the right pair of masses apart and stretch the left pair apart. As a result, the angle between the rods decreases until the spring counterbalances the tidal forces. This is the gradiometer's readout, its "readout angle."

If this gradiometer is flown rightward through the tidal pattern of Figure 24.6, its readout angle opens up above the oil-bearing region, and then closes down over the oil-free region. Gradiometers like this, but more sophisticated, are used by geologists to search for oil and also for mineral deposits.

NASA has flown a more sophisticated gradiometer called GRACE[1] (Figure 24.8) to map tidal fields everywhere above the Earth, and

Fig. 24.8. GRACE: Two satellites, which track each other with a beam of microwaves, are pushed together by blue tendex lines and stretched apart by red tendex lines. The tendex lines, from the Earth below, are not shown.

1 The Gravity Recovery and Climate Experiment, a joint US/German space mission launched in May 2002 and still collecting data in 2014.

watch slow changes of tidal gravity produced, for example, by the melting of ice sheets.

In my interpretation of *Interstellar*, most of the gravitational anomalies that Professor Brand's team measures are sudden and unexpected changes in the patterns of tendex lines above the Earth's surface, changes that occur for no obvious reason. The rocks and oil in the Earth's crust are not moving. The melting of ice sheets is much too slow to produce these quick changes. People see no new gravitating masses coming near the gradiometers. Nevertheless, the gradiometers report changing tidal patterns. Falling dust accumulates in radial lines. Cooper sees the coin plunge to the floor.

The members of Professor Brand's team monitor these changing patterns and eagerly record Cooper's observations. Their trove of data becomes grist for the Professor's quest to understand gravity, a quest that centers on the Professor's equation.

25

The Professor's Equation

I n *Interstellar,* the gravitational anomalies excite Professor Brand for two reasons. If he can discover their cause, that may trigger a revolution in our understanding of gravity, a revolution as great as Einstein's relativistic laws. More important: If he can figure out how to control the anomalies, that could enable NASA to lift large colonies of people off the dying Earth, and launch them toward a new home elsewhere in the universe.

For the Professor, the key to understanding and controlling the anomalies is an equation he has written on his blackboard (Figure 25.7, below). In the movie, he and Murph struggle to solve his equation.

Murph's and the Professor's Notebooks—and the Blackboard

Before filming began, two impressive Caltech physics students filled notebooks with calculations about the Professor's equation. Elena Murchikova filled a clean, new notebook with calculations by grown-up Murph, calculations written with elegant calligraphy. Keith Matthews filled a beat up, old notebook with calculations by Professor Brand, in the more sloppy handwriting common for old guys like the Professor and me.

In the movie, grown-up Murph (played by Jessica Chastain) discusses the math in her notebook with the Professor (played by Michael Caine). Murchikova, an expert in quantum gravity and cosmology, was on set to advise Chastain about her dialog and notebook, and things she was to write on the blackboard. It was startling to see these two brilliant and beautiful women from very different worlds, both with bright red hair, huddled together.

As for me, I filled Professor Brand's blackboard with diagrams and mathematics (Figure 25.8, below), including the Professor's equation—THE equation—at Christopher Nolan's request, of course. And I took great pleasure in talking with Michael Caine (Figure 25.1), who seemed to view me as a sort of prototype for the Professor he was playing. And great pleasure in watching Chris, a master craftsman, mold the scenes he was filming into precisely the form he wanted.

Some weeks before filming in the Professor's office, Chris and I went back and forth about what should be the nature of THE equation. (In Figure 1.2, back in Chapter 1, Chris is holding a sheaf of papers about the equation, which we are discussing.) Here's my long scientist's interpretation for what we wound up with—my extrapolation of the movie's story.

Source of the Anomalies—The Fifth Dimension

In my extrapolation, it does not take long for the Professor to convince himself that the anomalies are due to gravity from the fifth dimension. From the bulk. Why?

Fig. 25.1. Michael Caine (the Professor) and I, on set in the Professor's office.

The sudden changes in tidal gravity have no apparent source in our four-dimensional universe. For example, in my extrapolation the Professor's team sees the tidal gravity above an oil deposit switch, in just a few minutes, from the pattern we expect (top picture in Figure 25.2) to a radically different pattern (bottom picture). The oil has not moved. The rocks have not shifted. Nothing in our four-dimensional universe has changed except the tidal gravity.

These sudden changes *must* have a source. If the source is not in our universe, on our brane, then there is only one other place it can be, the Professor reasons: in the bulk.

In my extrapolation, the Professor can think of just three ways that something in the bulk could produce these anomalies, and the first two he quickly rejects:

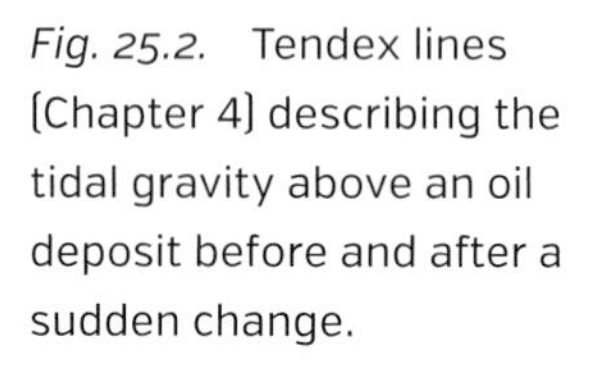

Fig. 25.2. Tendex lines (Chapter 4) describing the tidal gravity above an oil deposit before and after a sudden change.

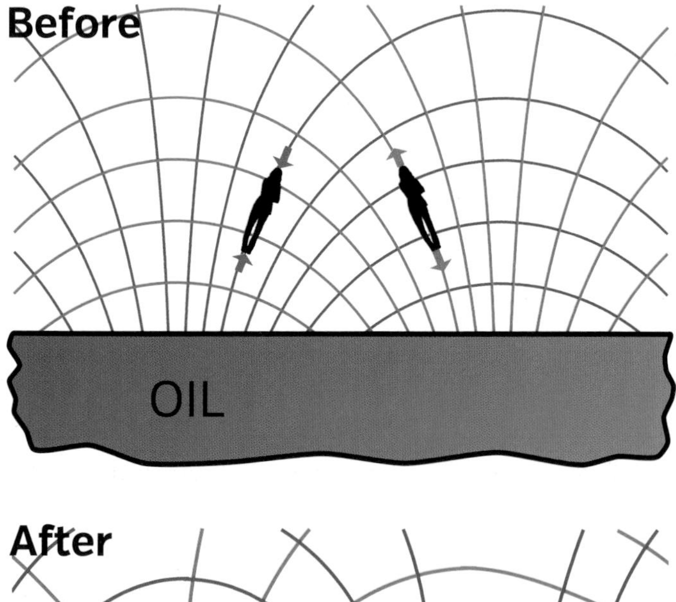

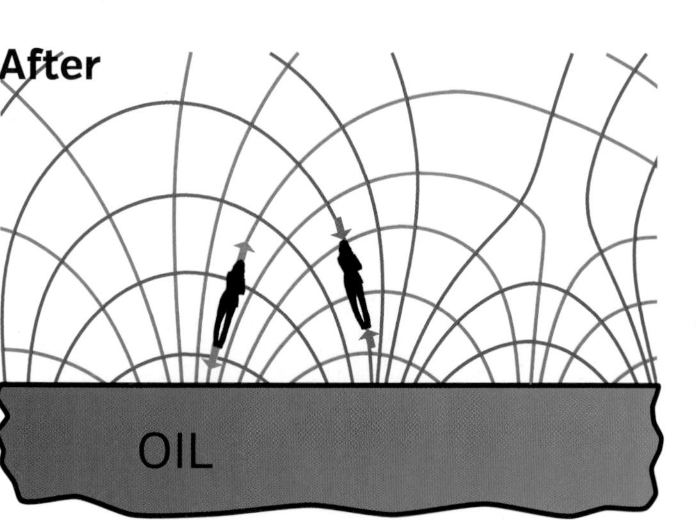

1. Some object in the bulk—perhaps even a living object, a bulk being—might come near our brane but not pass through it (upper right of Figure 25.3). The object's gravity reaches out through all the bulk's dimensions and so could reach into our brane. However, the AdS layer surrounding our brane (Chapter 23) would drive the object's tidal tendex lines parallel to our brane, allowing only a minuscule portion to reach our brane. So the Professor rejects this.
2. A bulk object, passing through our brane, could produce tidal gravity that changes as the bulk object moves (middle right of Figure 25.3). However, in my extrapolation most of the patterns of changing gravity that the Professor's team observed don't fit this explanation. The tendex lines tend to be more diffuse than those from a localized object. Some tidal anomalies might be from localized objects, but most must be something else.
3. Bulk fields passing through our brane could produce the changing tidal gravity (left side of Figure 25.3). This, the Professor concludes in my extrapolation, is the most likely explanation for most of the anomalies.

What is a "bulk field"? Physicists use the word *field* to mean something that extends out through space and exerts forces on things it encounters. We have already met several examples of fields that live in our universe, our brane: In Chapter 2, magnetic fields (collections of magnetic force lines), electric fields (collections of electric force lines), gravitational fields (collections of gravitational force lines); and in Chapter 4, tidal fields (collections of stretching and squeezing tendex lines).

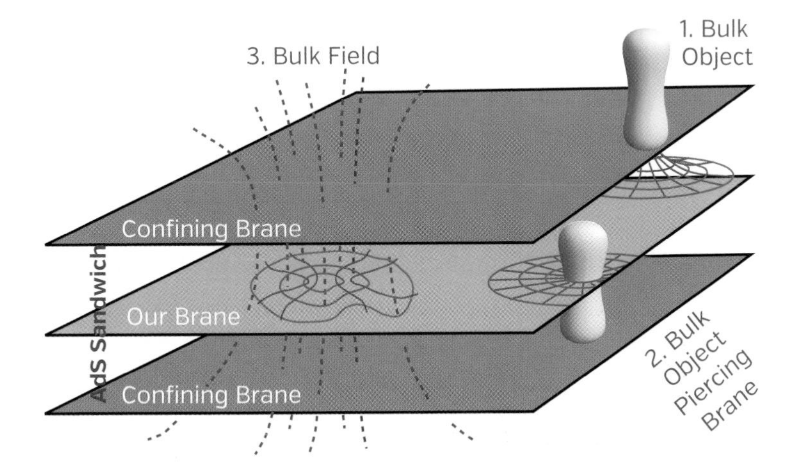

Fig. 25.3. Three ways that the bulk could produce the observed gravitational anomalies. The red and blue curves are tidal tendex lines produced by a bulk object or bulk field.

A bulk field is a collection of force lines that resides in the five-dimensional bulk. What kind of force lines, the Professor doesn't know, but he speculates; see below. Figure 25.3 shows a bulk field (dashed purple lines) passing through our brane. This bulk field generates tidal gravity in our brane (red and blue tendex lines). As the bulk field changes, its tidal gravity changes, resulting (the Professor thinks) in most of the observed anomalies.

But that isn't the only role of bulk fields, he suspects—in my extrapolation. They may also control the strength of the gravity produced by objects living in our brane, such as a rock or planet.

Bulk Fields Control the Strength of Gravity

The gravity of each little bit of matter in our brane is governed, to high accuracy, by Newton's inverse square law (Chapters 2 and 23): its gravitational pull is embodied in the formula $g = Gm/r^2$, where r is distance from that bit of matter, m is the mass of that bit of matter, and G is Newton's gravitational constant. This G controls the overall strength of the gravitational pull.

In Einstein's more accurate, relativistic version of the gravitational laws, the strength of gravity, and the strength of all the warping of space and time produced by matter, are also proportional to this G.

If there is no bulk—if the only thing that exists is our four-dimensional universe—then Einstein's relativistic laws say that G is absolutely constant. The same everywhere in space. Never changing in time.

But if the bulk *does* exist, then the relativistic laws allow this G to change. It *might*, the Professor speculates, be controlled by bulk fields. It *probably is* controlled by bulk fields, he thinks. That's the best explanation for one of the observed anomalies (Figure 25.4) in my extrapolation of the movie's story.

The strength of the Earth's gravitational pull varies slightly from place to place due to the varying density of the rocks, oil, oceans, and atmosphere. Earth-orbiting satellites have mapped this varying strength. As of 2014 the most accurate map is from the European Space Agency's satellite GOCE[1] (top half of Figure 25.4). In 2014, the

1 Gravity field and steady-state Ocean Circulation Explorer, GOCE.

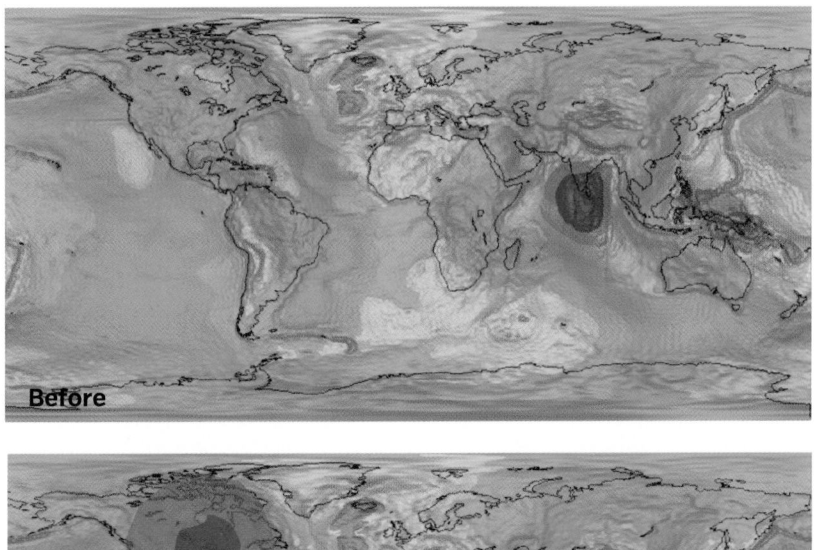

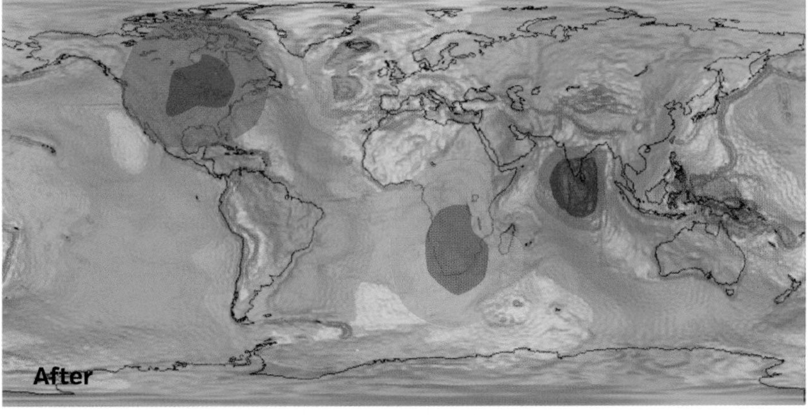

Fig. 25.4. Maps of the Earth's gravitational pull. *Top*: In 2014 as measured by the GOCE satellite. *Bottom*: After the sudden change in the era of anomalies.

Earth's gravity is weakest in southern India (blue spot) and strongest in Iceland and Indonesia (red spots).

In my extrapolation, this map did not change noticeably until anomalies started appearing. Then one day, quite suddenly, the Earth's gravitational pull in North America weakened a bit, and in South Africa it strengthened (bottom half of Figure 25.4).

Professor Brand tried to explain this as a change in the tidal forces produced by bulk fields, but had difficulty. The best explanation he could find is that the gravitational constant G increased inside the Earth, below South Africa, and decreased inside the Earth, below North America. Rock below South Africa was suddenly pulling more strongly; rock below North America was suddenly pulling more weakly! These changes must have been produced by some sort of bulk field that passes through our brane and controls G, he reasoned.

Bulk fields are not *just* the key to gravitational anomalies on Earth, Professor Brand believes (in my extrapolation). Bulk fields also play two other crucial roles: They hold the wormhole open, and they protect our universe from destruction.

Holding the Wormhole Open

The wormhole that connects our solar system to Gargantua's neighborhood, if left to its own devices, will pinch off (Figure 25.5). Our connection to Gargantua will be severed. This is the unequivocal conclusion of Einstein's relativistic laws (Chapter 14).

If there is *no* bulk, then the only way to hold the wormhole open is to thread it with exotic matter that repels gravitationally (Chapter 14). The dark energy that may accelerate our universe's expansion (Chapter 24) is probably not repulsive enough. In fact, it seems likely, in 2014, that the laws of quantum physics prevent even an exceedingly advanced civilization from ever collecting enough exotic matter to hold the wormhole open. And I imagine this conclusion is even more certain in Professor Brand's era.

But there is an alternative, the Professor realizes in my extrapolation of the movie's story. Bulk fields may do the job. They may hold the wormhole open. And since the Professor thinks the wormhole has been constructed and placed near Saturn by bulk beings, bulk fields holding it open seem natural to him.

Fig. 25.5. The wormhole. *Left*: Pinching off. *Right*: Held open by bulk fields.

Protecting Our Universe from Destruction

In order for gravity in our universe to obey Newton's inverse square law to high accuracy, our brane must be sandwiched between two confining branes with AdS warping between them (Chapter 23). However, the confining branes are filled with pressure[2] and prone to buckle, like a playing card pinched between two fingers (Figure 23.8). This is the unequivocal prediction of Einstein's relativistic laws, applied to the bulk and branes.

This buckling, if not counteracted, will make the confining branes collide with our brane—with our universe (Figure 25.6).[3] Our universe will be destroyed!

Obviously, our universe has not been destroyed, the Professor observes in my extrapolation. So something must prevent the confining branes from buckling. The only thing he can think of to do the job is bulk fields. Whenever a confining brane starts to bend, bulk fields must somehow exert a force on it, pushing it back into its proper, straight shape.

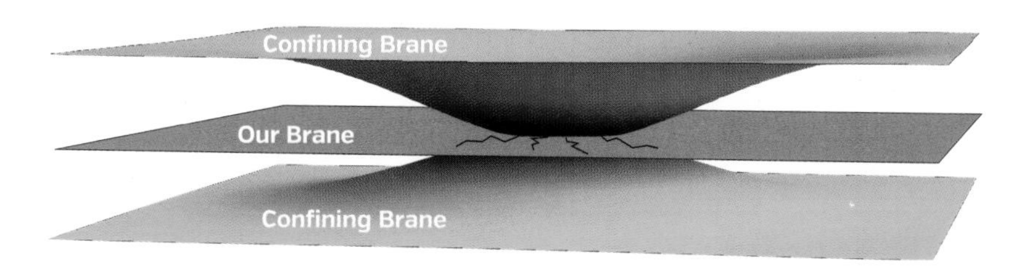

Fig. 25.6. Brane collision.

2 According to Einstein's relativistic laws the dark energy that (presumably) makes the expansion of our universe accelerate has a second effect: It produces an enormous tension in our brane, like the tension in a stretched rubber band or rubber sheet. And Einstein's laws also dictate that, in order for spacetime outside the AdS sandwich to be free of warping, as we desire, each confining brane must have internal pressure that is half as big as our own brane's internal tension. It is this pressure that is dangerous.

3 Or the buckling could make one or both branes spring outward, releasing the AdS layer and so destroying Newton's inverse square law and making the planets all fly away from the Sun—not quite so bad for our universe, but pretty miserable for humans.

The Professor's Equation, at Last!

The laws of physics are expressed in the language of mathematics. Before Cooper met Professor Brand (in my extrapolation of the movie's story), the Professor tried to build a mathematical description of the bulk fields and how they might generate anomalies, control our universe's gravitational constant G, hold the wormhole open, and protect our brane from collisions.

In creating this mathematics, the Professor was guided by the trove of observational data his team was collecting (Chapter 24), and by Einstein's relativistic laws of physics in five dimensions.

The Professor embodied all his insights in a single equation, THE equation, which he wrote on one of the sixteen blackboards in his office (Figure 25.7).[4] Cooper sees the equation on his first visit to NASA, and the equation is still there thirty years later, when Murph has grown up to become a brilliant physicist in her own right, and is helping the Professor try to solve it.

Fig. 25.7. Professor Brand's equation.

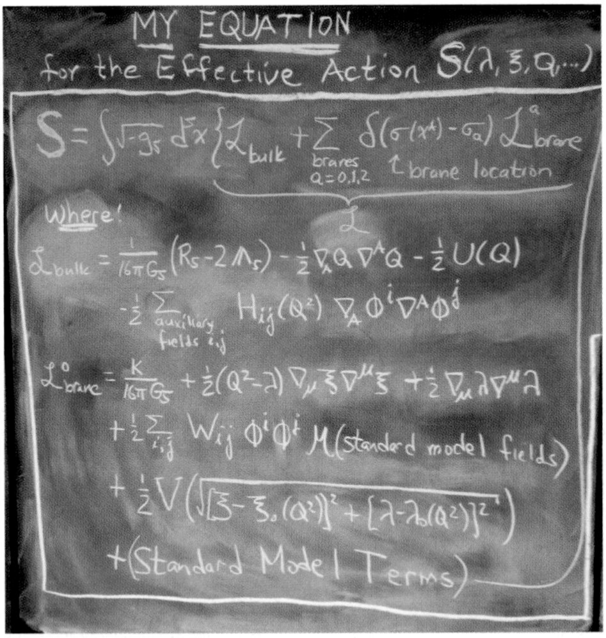

This equation is called an "Action." There is a well-known (to physicists) mathematical procedure to begin with such an Action, and from it deduce *all* the nonquantum physical laws. The Professor's equation, in effect, is the mother of all nonquantum laws. But for it to give birth to the *right* laws—the laws that predict *correctly* how the anomalies are produced, how the wormhole is held open, how G is controlled, and how our universe is protected—the equation must have precisely the correct mathematical form. The Professor doesn't know the correct form. He is guessing. His is an educated guess, but a guess nevertheless.

His equation contains lots of guessing:

4 The meanings of the various symbols in the equation are spelled out on the Professor's other fifteen blackboards, along with other information about the equation, all of which I ghost-wrote for the movie's filming. You can see photographs of all sixteen blackboards on this book's page at Interstellar.withgoogle.com.

guesses for the things called "$U(Q)$, $H_{ij}(Q^2)$, W_{ij}, and $\mathcal{M}$(standard model fields)" on his blackboard (Figure 25.7). In effect, these are guesses for the nature of the bulk fields' force lines, and how they influence our brane, and how fields in our brane influence them. (For more explanation see *Some Technical Notes* at the end of this book.)

Fig. 25.8. I ghost-write iterative guesses on the Professor's blackboard.

When the Professor and his team speak of "solving his equation," in my extrapolation they mean two things. First, figure out the *right* forms for all these things they are guessing: "$U(Q)$, $H_{ij}(Q^2)$, W_{ij}, and $\mathcal{M}$(standard model fields)." Second (following the well-known procedure), deduce, from his equation, everything he wants to know about our universe, about the anomalies, and most important, about how to control the anomalies so as to lift colonies off the Earth.

When characters in the movie speak of "solving gravity," they mean the same thing.

In the movie, when the Professor is very old, we see him and grown-up Murph trying to solve his equation by iterations. On a blackboard, they make a list of guesses for the unknown things (guesses that I wrote on the board just before the scene was filmed; Figures 25.8 and 25.9). Then, in my extrapolation, Murph

Fig. 25.9. Murph contemplates the list of iterative guesses. *[From* Interstellar, *used courtesy of Warner Bros. Entertainment Inc.]*

inserts each guess into a huge computer program that they've written. The program computes the physical laws for that guess, and those laws' predictions for how the gravitational anomalies behave.

In my extrapolation, none of the guesses predicts anomalies that look anything like the observations. But in the movie, the Professor and Murph keep trying. They keep iterating: making a guess, computing the consequences, abandoning the guess, and going on to the next guess, one guess after another after another after another, until exhaustion sets in. Then they begin again the next day.

A bit later in the movie, when the Professor is on his deathbed, he confesses to Murph: "I lied, Murph. I lied to you." It is a poignant scene. Murph infers that he knew something was wrong with his equation, knew from the outset. And Dr. Mann tells the Professor's daughter as much in an equally poignant scene on Mann's planet.

But, in fact—Murph realizes, soon after the Professor's death—"His solution was correct. He'd had it for years. It's half the answer." The other half can be found inside a black hole. In a black hole's singularity.

26

Singularities and Quantum Gravity

In *Interstellar* Cooper and TARS seek quantum data inside Gargantua, data that could help the Professor solve his equation and lift humanity off Earth. The data, they believe, must reside inside a singularity that inhabits Gargantua's core—a "gentle" singularity, Romilly predicts. What are the quantum data? How could they help the Professor? And what is a gentle singularity?

The Primacy of Quantum Laws

Ⓣ

Our universe is fundamentally quantum. By this I mean that everything fluctuates randomly, at least a little bit. Everything!

When we use high-precision instruments to look at tiny things, we see big fluctuations. The location of an electron inside an atom fluctuates so rapidly and so randomly, that we can't know where the electron is at any moment of time. The fluctuations are as big as the atom itself. That's why the quantum laws of physics deal with *probabilities* for where the electron is and not with its actual location (Figure 26.1).

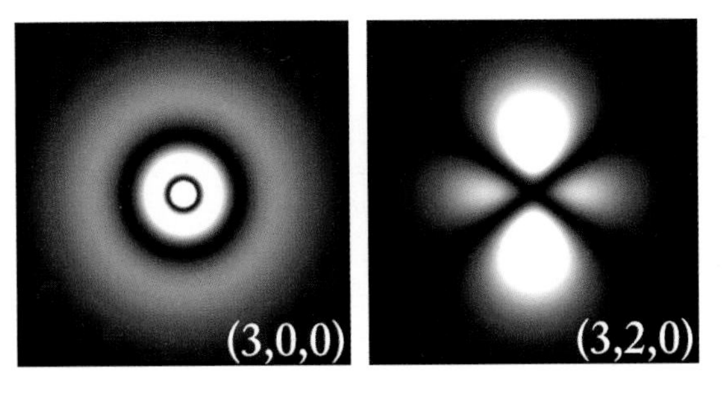

Fig. 26.1. Probability for electron's location inside two different hydrogen atoms. The probability is big in the white regions, smaller in the red, and very small in the black. The numbers (3,0,0) and (3,2,0) are the names of the two atoms' probability pictures.

When we use instruments to look at big things, we also see fluctuations, if our instruments are precise enough. But the fluctuations of big things are minuscule. In the LIGO gravitational wave detectors (Chapter 16), laser beams monitor the locations of hanging mirrors that weigh 40 kilograms (90 pounds).[1] Those locations fluctuate randomly, but by amounts far less than the size of an atom: *one ten-billionth of an atom's size,* in fact (Figure 26.2). Nevertheless, LIGO's laser beams will see those fluctuations a few years from now. (LIGO's design prevents those random fluctuations from getting in the way of measuring gravitational waves. My students and I helped make sure of this.)

Because objects of human size and larger have only minuscule quantum fluctuations, physicists almost always ignore those fluctuations. Discarding the fluctuations, in our mathematics, simplifies the laws of physics.

If we begin with the ordinary quantum laws that ignore gravity and then discard the fluctuations, we obtain the Newtonian laws of physics—the laws used for the past few centuries to describe planets, stars, bridges, and marbles. See Chapter 3.

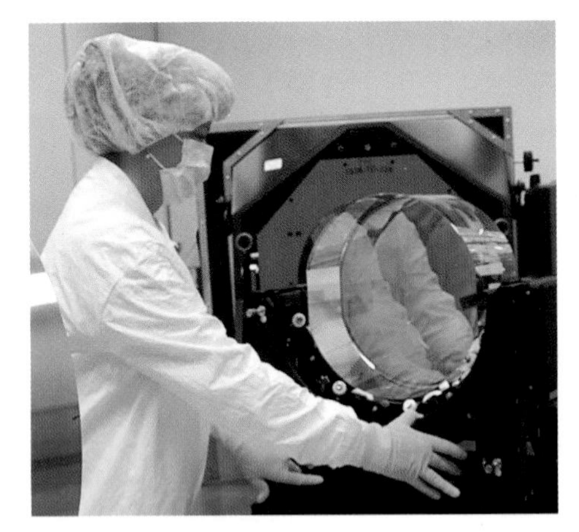

Fig. 26.2. A 40-kilogram mirror being prepared for installation in LIGO. Its location fluctuates, quantum mechanically, very, very slightly: one ten-billionth the diameter of an atom.

If we begin with the ill-understood laws of quantum gravity and then discard the fluctuations, we must obtain Einstein's well-understood relativistic laws of physics. The fluctuations we discard are, for example, a froth of fluctuating, exquisitely tiny wormholes ("quantum foam" that pervades all of space; Figure 26.3 and Chapter 14).[2] With the fluctuations gone, Einstein's laws describe the precise warping of space and time around black holes, and the precise slowing of time on Earth.

This is all the preamble to a punch line: *If Professor Brand could dis-*

1 More precisely, the locations of the mirrors' centers of mass.

2 In 1955, John Wheeler pointed out the likely existence of quantum foam, with wormhole sizes 10^{-35} meters: 10 trillion trillion times smaller than an atom; the so-called *Planck length.*

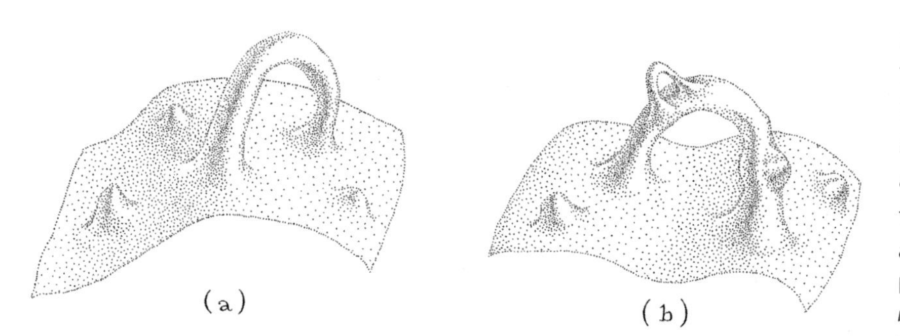

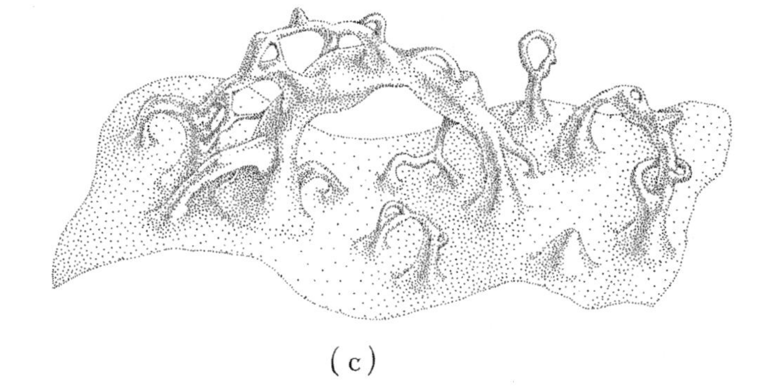

Fig. 26.3. Quantum foam. There is some probability [say, 0.4] that the foam will have the upper left shape, another probability [say, 0.5] for the upper right shape, and another [say, 0.1] for the lower shape. *[Drawing by Matt Zimet based on a sketch by me; from my book* Black Holes & Time Warps: Einstein's Outrageous Legacy.*]*

cover the quantum gravity laws for the bulk as well as our brane, then by discarding those laws' fluctuations, he could deduce the precise form of his equation (Chapter 25). And that precise form would tell him the origin of the gravitational anomalies and how to control the anomalies—how to employ them (he hopes) to lift colonies off Earth.

In my extrapolation of the movie, the Professor knows this. And he also knows a place where the quantum gravity laws can be learned: inside *singularities.*

Singularities: The Domain of Quantum Gravity

Ⓣ

The beginning of a singularity is a place where the warping of space and time grows without bound. Where space warps and time warps become infinitely strong.

If we think of our universe's warped space as like the undulating surface of the ocean, then the beginning of a singularity is like the tip of a wave that is about to break, and the interior of the singularity is

Fig. 26.4. A singularity at the tip of an ocean wave that is about to break.

like the froth after it breaks (Figure 26.4). The smooth wave, before it breaks, is governed by smooth laws of physics, analogs of Einstein's relativistic laws. The froth after it breaks requires laws capable of dealing with frothing water, analogs of the laws of quantum gravity with their quantum foam.

Singularities inhabit the cores of black holes. Einstein's relativistic laws predict them unequivocally, even though those laws can't tell us what happens inside the singularities. For that, we need the quantum gravity laws.

In 1962 I moved from Caltech (my undergraduate school) to Princeton University, to study for a PhD in physics. I chose Princeton because John Wheeler taught there. Wheeler was that era's most creative genius, when it comes to Einstein's relativistic laws. I wanted to learn from him.

One September day, with trepidation I knocked on the door of Professor Wheeler's office. It would be my first meeting with the great man. He greeted me with a warm smile, ushered me in, and immediately—as though I were an esteemed colleague, not a total novice—began discussing the mysteries of the implosions of stars. Implosions that produce black holes with singularities in their cores. These singularities, he asserted, "are a place in which the fiery marriage of Einstein's relativistic laws with the quantum laws is consummated." The fruits of that marriage, the laws of quantum gravity, come into full blossom in singularities, Wheeler asserted. If we could understand singularities, we would learn the laws of quantum gravity. Singularities are a rosetta stone for deciphering quantum gravity.

Fig. 26.5. John Wheeler in 1971, lecturing about singularities, black holes, and the universe.

From that private lecture, I emerged a convert. From Wheeler's public lectures and writings, many other physicists emerged as converts and embarked on a quest to understand singularities and their quantum gravity laws. That quest continues today. That quest produced superstring theory, which in turn led to a belief that our universe must be a brane residing in a higher dimensional bulk (Chapter 21).

Naked Singularities?

(EG)

It would be fabulous if we could find or make a singularity *outside* a black hole. A singularity not hidden beneath a black hole's event horizon. A *naked singularity*. Then in *Interstellar* the Professor's task could be easy. He might extract the crucial quantum data from a naked singularity in his NASA lab.

In 1991, John Preskill and I made a bet about naked singularities with our friend Stephen Hawking. Preskill, a Caltech professor, is one of the world's great experts on quantum information. Stephen is the "wheelchair guy" who appears on *Star Trek, The Simpsons,* and *The Big Bang Theory.* He also happens to be one of the greatest geniuses of our era. John and I bet the laws of physics permit naked singularities. Stephen bet they are forbidden (Figure 26.6).

None of us thought the bet would be resolved quickly, but it was. Just five years later Matthew Choptuik, a postdoctoral student at the University of Texas, carried out a simulation on a supercomputer

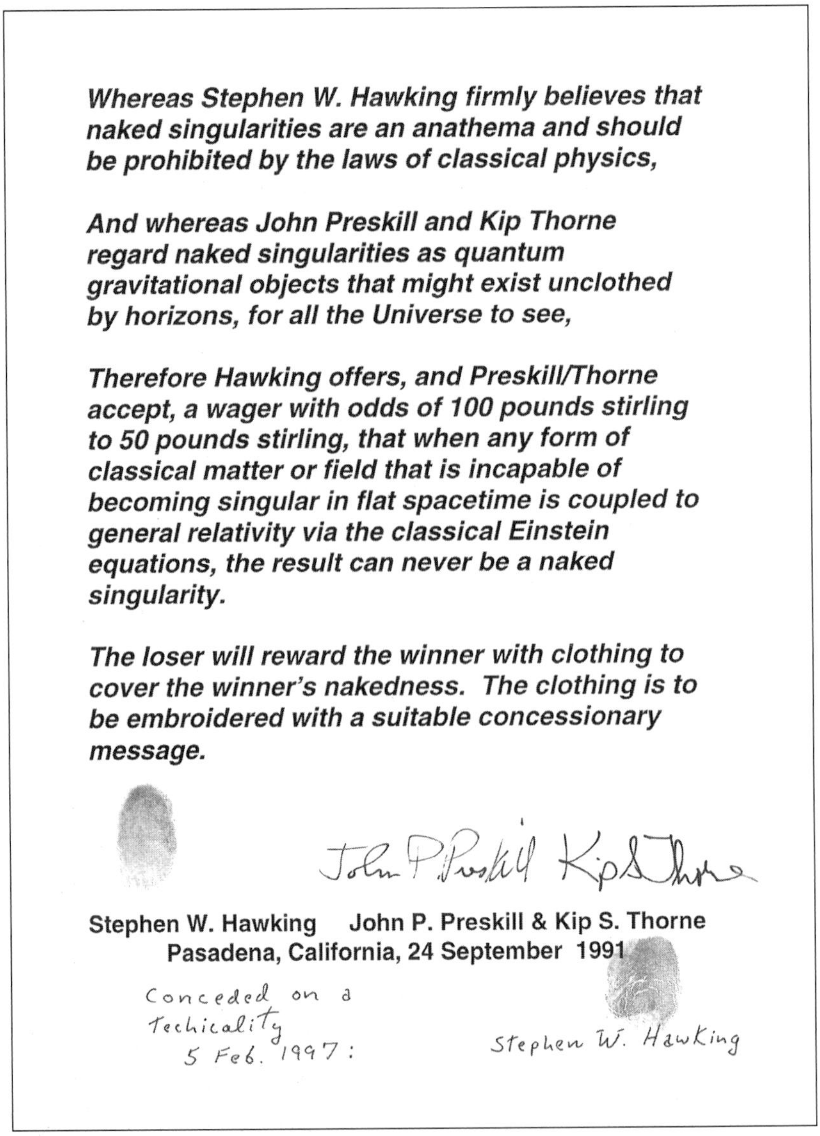

Fig. 26.6. Our bet about naked singularities.

Whereas Stephen W. Hawking firmly believes that naked singularities are an anathema and should be prohibited by the laws of classical physics,

And whereas John Preskill and Kip Thorne regard naked singularities as quantum gravitational objects that might exist unclothed by horizons, for all the Universe to see,

Therefore Hawking offers, and Preskill/Thorne accept, a wager with odds of 100 pounds stirling to 50 pounds stirling, that when any form of classical matter or field that is incapable of becoming singular in flat spacetime is coupled to general relativity via the classical Einstein equations, the result can never be a naked singularity.

The loser will reward the winner with clothing to cover the winner's nakedness. The clothing is to be embroidered with a suitable concessionary message.

John P. Preskill Kip S. Thorne

Stephen W. Hawking John P. Preskill & Kip S. Thorne
Pasadena, California, 24 September 1991

Conceded on a techicality 5 Feb. 1997: Stephen W. Hawking

that he hoped would reveal new, unexpected features of the laws of physics; and he hit the jackpot. What he simulated was the implosion of a gravitational wave.[3] When the imploding wave was weak, it imploded and then disbursed. When it was strong, the wave imploded and formed a black hole. When its strength was very pre-

3 The thing he simulated was actually something called a scalar wave, but that is an irrelevant technicality. A few years later Andrew Abrahams and Chuck Evans at the University of North Carolina repeated Choptuik's simulations using a gravitational wave and got the same result: a naked singularity.

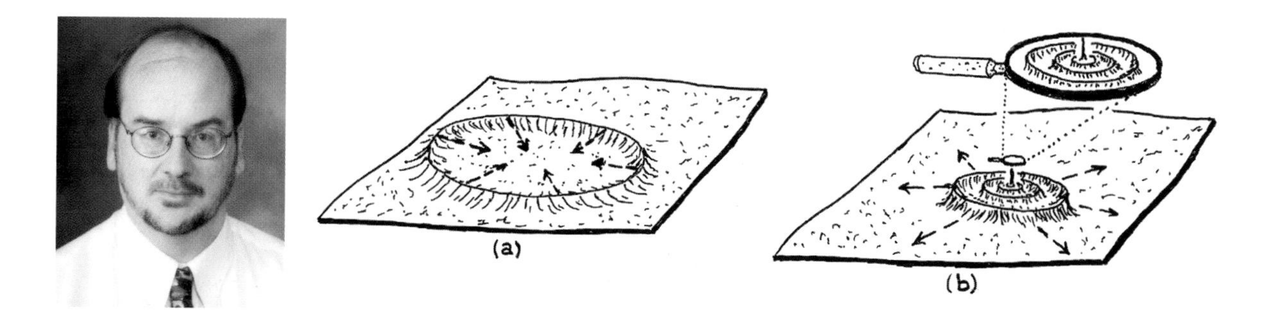

cisely "tuned" to an intermediate strength, the wave created a sort of boiling in the shapes of space and time. The boiling produced outgoing gravitational waves with shorter and shorter wavelengths. It also left behind, at the end, an infinitesimally tiny naked singularity (Figure 26.7).

Fig. 26.7. *Left*: Matthew Choptuik. *Middle*: An imploding gravitational wave. *Right*: The boiling produced by the wave, and the naked singularity at the center of the magnifying glass.

Now, such a singularity can never occur in nature. The required tuning is not a natural thing. But an exceedingly advanced civilization could produce such a singularity artificially by precisely tuning a wave's implosion, and then could try to extract the laws of quantum gravity from the singularity's behavior.

Upon seeing Choptuik's simulation, Stephen conceded our bet—"on a technicality," he said (bottom of Figure 26.6). He thought precise tuning unfair. He wanted to know whether naked singularities can occur *naturally*, so we renewed our bet with a new wording that the singularity must arise without any need for precise tuning. Nevertheless, Stephen's concession, in a very public venue (Figure 26.8) was a big deal. It made the front page of the *New York Times*.

Fig. 26.8. Hawking conceding to Preskill and Thorne at a 1997 Caltech lecture by Hawking.

Despite our renewed bet, I doubt that naked singularities do exist in our universe. In *Interstellar*, Dr. Mann firmly asserts that "the laws of nature prohibit a naked singularity," and Professor Brand never even mentions that possibility. Instead, the Professor focuses on singularities inside black holes. Those, he thinks, are the only hope for learning the laws of quantum gravity.

The BKL Singularity Inside a Black Hole

(EG)

In Wheeler's era (the 1960s), we thought of a singularity inside a black hole as like a sharp point. A point that squeezes matter until the matter becomes infinitely dense and is destroyed. That's how, until now in this book, I have depicted a black hole's singularity (Figure 26.9, for example).

Since Wheeler's era, mathematical calculations with Einstein's laws have taught us that these pointy singularities are unstable. To create them inside a black hole requires precise tuning. When perturbed ever so slightly, for example by something falling in, they change enormously. Change into what?

Three Russian physicists—Vladimir Belinsky, Isaac Khalatnikov,

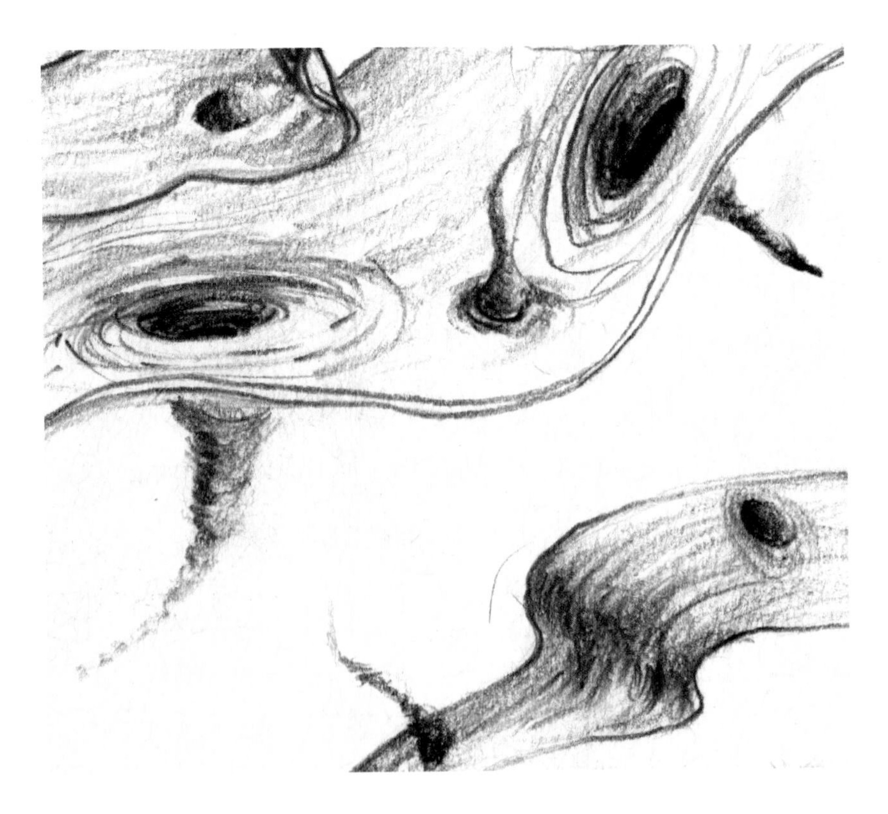

Fig. 26.9. Lia Halloran's fanciful drawing of several black holes with singularities at their pointy tips. *[A segment out of Fig. 4.5.]*

and Eugene Lifshitz—used long, complicated calculations to guess the answer, in 1971. And in the 2000s, when computer simulations became sufficiently advanced, their guess was confirmed by David Garfinkle at Oakland University. The resulting, stable singularities now carry the name BKL in honor of Belinsky, Khalatnikov, and Lifshitz.

A BKL singularity is chaotic. Highly chaotic. And lethal. Highly lethal.

In Figure 26.10, I depict the warping of space outside and inside a fast-spinning black hole. The BKL singularity is at the bottom. If you fall into this black hole, its interior at first is smooth and perhaps pleasant. But as you near the singularity, the space around you begins to stretch and squeeze in a chaotic pattern. And tidal forces begin to stretch and squeeze you, chaotically. The stretch and squeeze are gentle at first, but quickly they become strong, then ultrastrong. Your flesh and bones are pummeled and give way. Then the atoms of which your body was made are pummeled and give way—distorted beyond recognition.

All this and its chaotic pattern are described by Einstein's relativistic laws. It is this that the Russians, B, K, and L, predicted. What they could not predict, what nobody can predict today, is the fate of your atoms and subatomic particles when the chaotic pummeling grows to an infinite crescendo. Only the laws of quantum gravity know their fate. But you, yourself, are long since dead, with no possibility to retrieve the quantum data and escape.

I labeled this section (EG) for educated guess, because we are *not* absolutely certain that the singularity inside a black hole's core is a BKL one. BKL singularities are surely allowed by Einstein's relativistic laws. Garfinkle confirmed it by computer simulations. But more sophisticated simulations are needed to confirm that the BKL patterns of humongous stretch and squeeze do actually occur in the core of a black hole. I'm almost sure the result of those simulations will be "yes, they do occur." But I'm not completely certain.

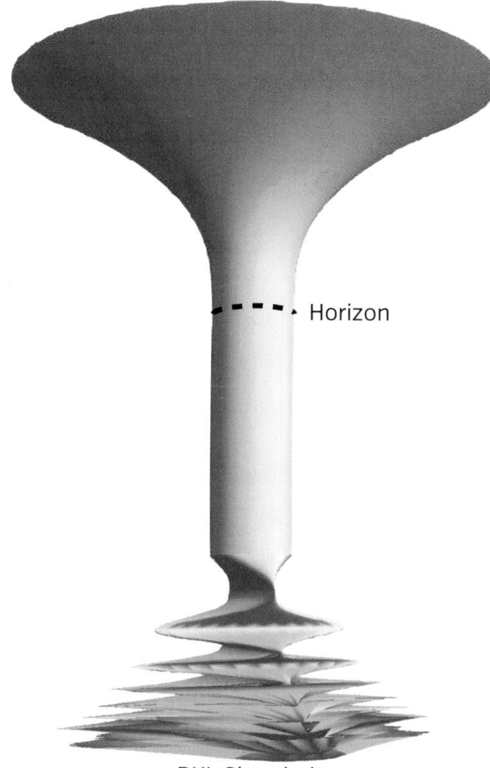

Fig. 26.10. The warped space of a fast-spinning black hole such as Gargantua, with the BKL singularity at the bottom. The chaotic stretch and squeeze near the singularity are depicted heuristically, not precisely.

A Black Hole's Infalling and Outflying Singularities

(EG)

My physicist colleagues and I were pretty sure in the 1980s, as an educated guess, that there is just one singularity inside a black hole, and it's a BKL singularity. We were wrong.

In 1991 Eric Poisson and Werner Israel at the University of Alberta, Canada, working with the mathematics of Einstein's laws, discovered a second singularity. This one grows with time as the black hole ages. It's caused by extreme slowing of time inside the black hole.

If you fall into a spinning black hole such as Gargantua, lots of other stuff inevitably will fall in after you: gas, dust, light, gravitational waves, and so forth. This stuff may take millions or billions of years to enter the hole as seen by me, watching from outside. But as seen by you, now inside the hole, it may take only a few seconds or less, due to the extreme slowing of your time compared with mine. As a result, as seen by you this stuff all piles up in a thin sheet, falling inward toward you at the speed of light, or nearly the speed of light. This sheet generates intense tidal forces that distort space and will distort you, if the sheet hits you.

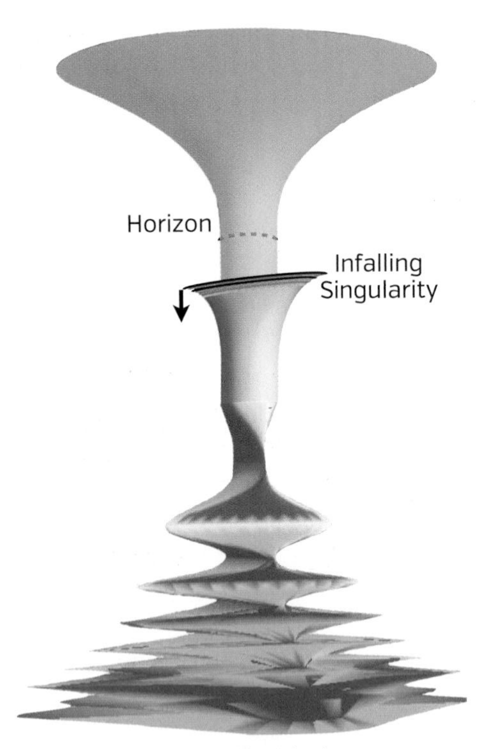

Fig. 26.11. The infalling singularity, created by stuff that falls into the black hole after you. The stuff is epitomized by alternating layers of black, red, gray, and orange.

The tidal forces grow to become infinite. The result is an "infalling singularity" (Figure 26.11),[4] governed by the laws of quantum gravity. However, the tidal forces grow so swiftly (Poisson and Israel deduced) that, if they hit you, they will have deformed you by only a finite amount at the moment you reach the singularity. This is explained in Figure 26.12, which plots your net stretch along the up-down direction and squeeze along the north-south and east-west directions, as time passes. When you hit the singularity, your net stretch and squeeze are finite, but the rates at which you are being

4 Israel and Poisson gave this singularity the name *mass inflation singularity*, and that is the name that physicists have used ever since. I prefer *infalling singularity* and use that name in this book.

stretched and squeezed (the slopes of the black curves) are infinite. Those infinite rates are the infinite tidal forces, signaling the singularity.

Because your body has been stretched and squeezed by only a finite net amount, when you reach the singularity, it is conceivable you might survive. (Conceivable but unlikely, I think.) In this sense, the infalling singularity is far more "gentle" than the BKL singularity. If you do survive, what happens next is known only to the laws of quantum gravity.

In the 1990s and 2000s, we physicists thought this was the whole story: A BKL singularity, created when the black hole is born. And an infalling singularity that grows afterward. That's all.

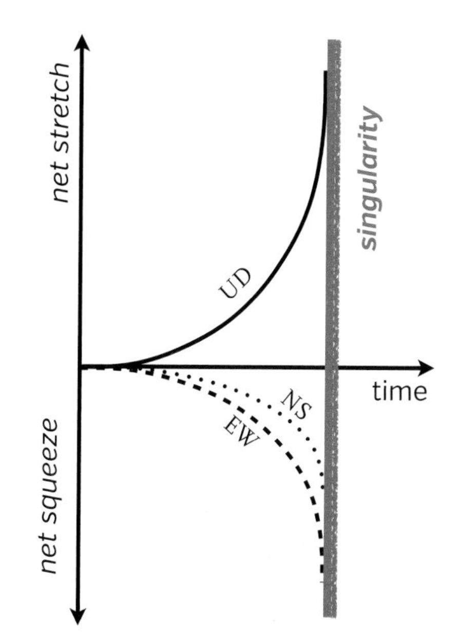

Fig. 26.12. Your net stretch and squeeze, as time passes, when the infalling singularity descends on you.

Then in late 2012, while Christopher Nolan was negotiating to rewrite and direct *Interstellar*, a third singularity was discovered by Donald Marolf (University of California at Santa Barbara) and Amos Ori (The Technion, in Haifa, Israel). It was discovered, of course, via an in-depth study of Einstein's relativistic laws and not via astronomical observations.

In retrospect, this singularity should have been obvious. It is an outflying singularity that grows as the black hole ages, just like the infalling singularity grows. It is produced by stuff (gas, dust, light, gravitational waves, etc.) that fell into the black hole *before* you fell in; Figure 26.13. A tiny fraction of that stuff is scattered back upward toward you, scattered by the hole's warpage of space and of time, much like sunlight scattered off a curved, smooth ocean wave, which brings us an image of the wave.

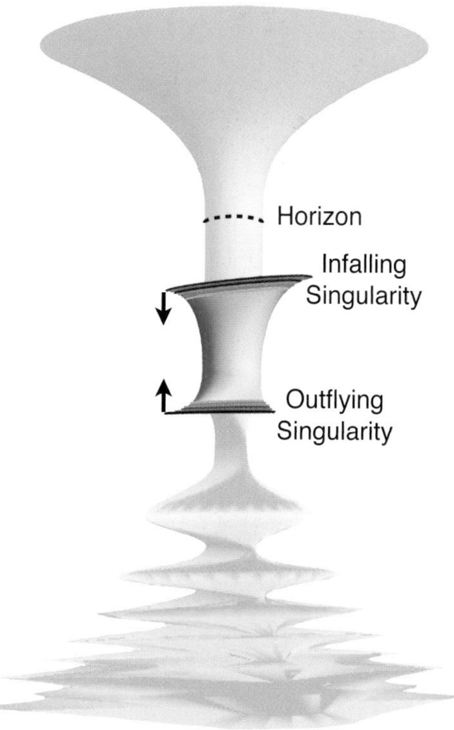

Fig. 26.13. The outflying singularity, created by backscattered stuff that fell into the black hole before you; and the infalling singularity, created by stuff that falls in after you. You are sandwiched between them. Shown dimmed out is the exterior of the black hole and the BKL singularity, with which you can no longer have contact because they are beyond the singularities that sandwich you.

The upscattered stuff gets compressed, by the black hole's extreme slowing of time, into a thin layer rather like a sonic boom (a "shock front"). The stuff's gravity produces tidal forces that grow infinitely strong and thence become an *outflying singularity*. But as for the infalling singularity, so also for this outflying one, the tidal forces are gentle: They grow so quickly, so suddenly, that, if you encounter one, your net distortion is finite, not infinite, at the moment you hit the singularity.

In *Interstellar*, Romilly tells Cooper about these gentle singularities: "I have a suggestion for your return journey [from Mann's planet]. Have one last crack at the black hole. Gargantua's an older, spinning black hole. [It has] what we call a gentle singularity." "Gentle?" Cooper asks. "They're hardly gentle, but their tidal gravity is quick enough that something crossing the horizon fast might survive." Cooper, lured by this conversation and the quest for quantum data, later plunges into Gargantua (Chapter 28). It's a brave plunge. He can't know in advance whether he'll survive. Only the laws of quantum gravity know for sure. Or the bulk beings . . .

We've now laid the extreme-physics foundations for *Interstellar*'s climactic scenes, so let's turn to the climax.

VII

CLIMAX

27

The Volcano's Rim

Ⓣ

Late in *Interstellar,* Cooper has just dragged the *Endurance* out of its death spiral at Mann's planet and feels a great sense of relief when the robot CASE says to him: "We're heading into Gargantua's pull."

Cooper makes a quick decision: "The navigation mainframe's destroyed and we don't have enough life support to make it back to Earth. But we might scrape to Edmunds' planet." "What about fuel?" Amelia Brand asks. "Not enough," Cooper responds. "Let Gargantua suck us right to [near] the horizon, then a powered slingshot around to launch us at Edmunds' planet." "Manually?" "That's what I'm here for. I'll take us just inside the critical orbit."

Within minutes they are at the critical orbit and all hell breaks loose.

In this chapter, I describe my scientist's interpretation of this.

Tidal Gravity: Breaking the *Endurance* Away from Mann's Planet

In my interpretation Mann's planet is on a highly elongated orbit (Chapter 19). When the *Endurance* arrived at the planet, it was rather far from

Gargantua but zooming inward. The *Endurance*'s explosion (Chapter 20) occurred when the planet was nearing the black hole (Figure 27.1).

Cooper rescues the *Endurance* after the explosion and lifts it upward, away from the planet. In my interpretation, he lifts the *Endurance* high enough for Gargantua's huge tidal forces to pry it away from the planet, sending it on a separate trajectory (Figure 27.2).

Centrifugal forces fling Mann's planet outward on its next distant excursion, while the *Endurance* heads onto the critical orbit.[1]

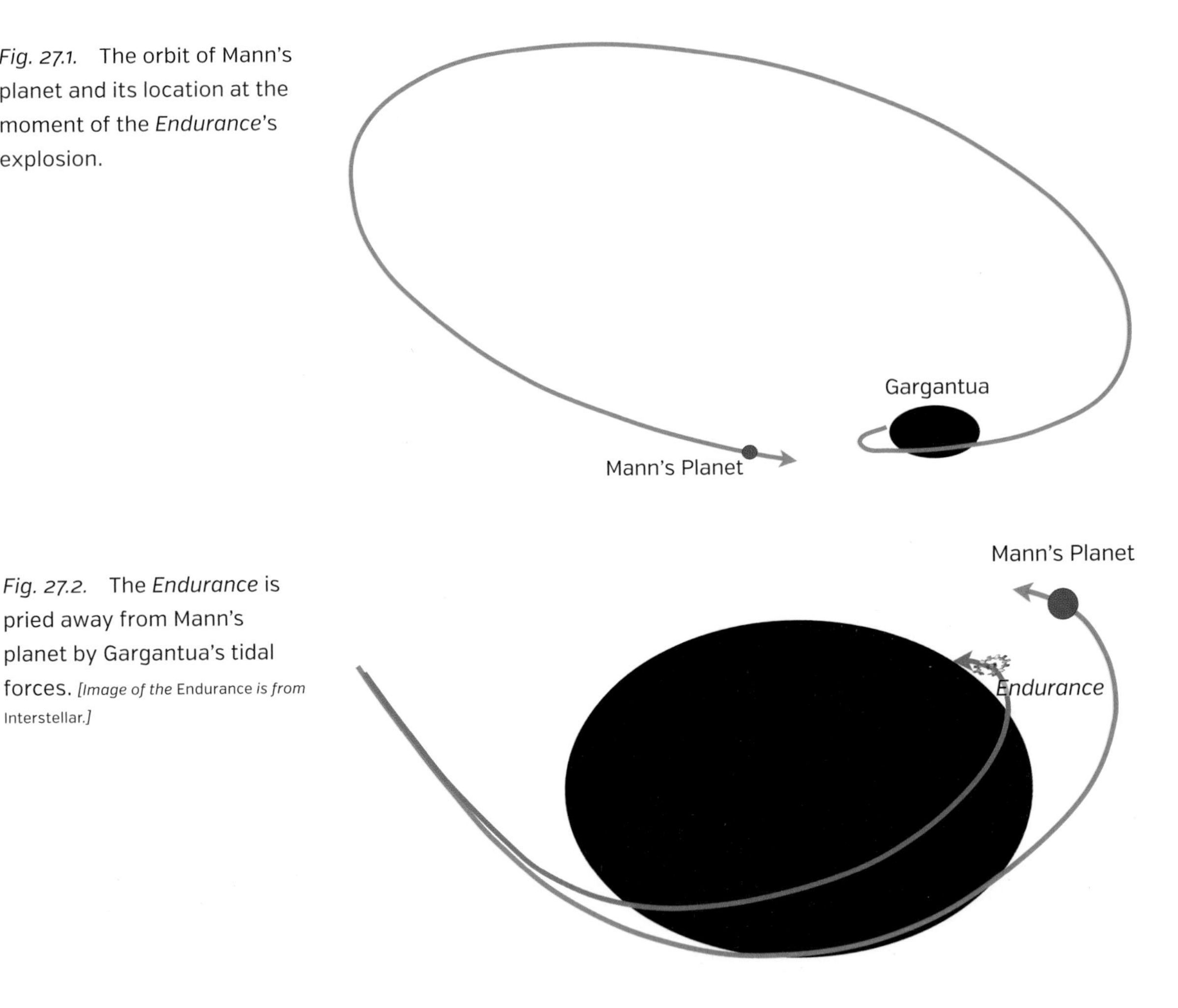

Fig. 27.1. The orbit of Mann's planet and its location at the moment of the *Endurance*'s explosion.

Fig. 27.2. The *Endurance* is pried away from Mann's planet by Gargantua's tidal forces. *[Image of the* Endurance *is from* Interstellar.*]*

1 This big difference is due to the *Endurance*'s having slightly less angular momentum than Mann's planet, after tidal forces have done their thing. In Figure 27.3 the *Endurance* climbs up onto the volcano's rim, but Mann's planet does not quite make it up to the rim; it spirals back down the volcano's side (centrifugal forces push it outward) and then up the gravitational energy surface, away from Gargantua.

The Critical Orbit and the Volcano Analogy

I discuss the critical orbit using a different type of picture than I've used before: Figure 27.3. I first describe this picture heuristically, and then I explain it in physicists' language.

Think of the surface in Figure 27.3 as that of a smooth, granite sculpture sitting on the floor in your home. It sinks down to a deep moat that surrounds a sculpted volcano.

The *Endurance*, after being pried away from Mann's planet, is like a tiny marble that rolls freely on this granite surface. As it rolls inward toward the moat, the marble picks up speed, because of the surface's downward slope. It then rolls up the volcano's side, slowing as it goes, and arrives on the volcano's rim with some residual circumferential motion. And it then rolls around and around on the rim, delicately and unstably balanced between falling inward, into the volcano, and falling back outward and down to the moat.

The volcano's interior is Gargantua, and the volcano's rim is the critical orbit, from which the *Endurance* launches toward Edmunds' planet.

The Meaning of the Volcano: Gravitational and Circumferential Energy

To explain the volcano's meaning—how it relates to the laws of physics—I have to get a bit technical.

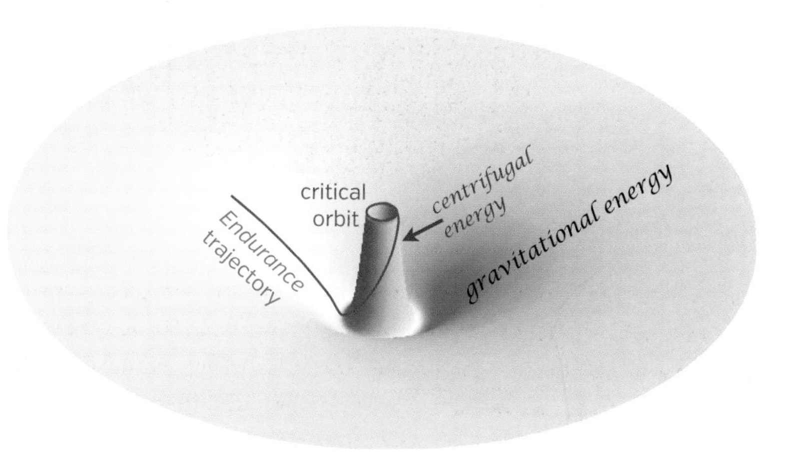

Fig. 27.3. The *Endurance*'s trajectory on a volcano-like surface that represents its gravitational and centrifugal energies.

For the sake of simplicity, let's pretend the *Endurance* is moving in Gargantua's equatorial plane. (For the *Endurance*'s nonequatorial trajectory the ideas are the same but because the black hole is not spherical, the details are more complicated.) The volcano analogy neatly encapsulates the true physics of the critical orbit and Gargantua's trajectory. To explain how, I need two physics concepts: the *Endurance*'s *angular momentum,* and its *energy*.

After tidal forces pry it apart from Mann's planet, the *Endurance* has a certain amount of angular momentum (its circumferential speed around Gargantua times its distance from Gargantua). The relativistic laws tell us that this angular momentum remains fixed (conserved) along the *Endurance*'s trajectory; see Chapter 10. This means that, as the *Endurance* plunges toward Gargantua, with its distance from Gargantua decreasing, its circumferential speed increases. This is similar to an ice-skater, whose whirling speed increases when she moves her arms in (Figure 27.4).

The *Endurance* heads toward Gargantua with a certain amount of energy, which like its angular momentum remains constant along its trajectory. This energy consists of three parts: the *Endurance*'s *gravitational energy,* which gets more and more negative as the *Endurance* plunges toward Gargantua; its *centrifugal energy* (its energy of circumferential motion around Gargantua), which increases as the *Endurance* plunges because the circumferential motion is speeding up; and its *radial kinetic energy* (its energy of motion toward Gargantua).

The surface in Figure 27.3 is the *Endurance*'s gravitational energy plus its centrifugal energy plotted vertically, and location in Gargantua's equatorial plane plotted horizontally. Wherever the surface dips downward, the *Endurance*'s gravitational plus centrifugal energy decreases, so its radial kinetic energy must increase (since the total energy is unchanged); its radial motion must speed up. This is precisely what happens in our intuitive, volcano analogy.

Fig. 27.4. Ice skater.

Outside the moat of Figure 27.3, the surface's height is controlled by the *Endurance*'s negative gravitational energy (see the "gravitational energy" label on the figure). By comparison, there the positive centrifugal energy is unimportant. On the outer edge of the volcano, by contrast, the height is

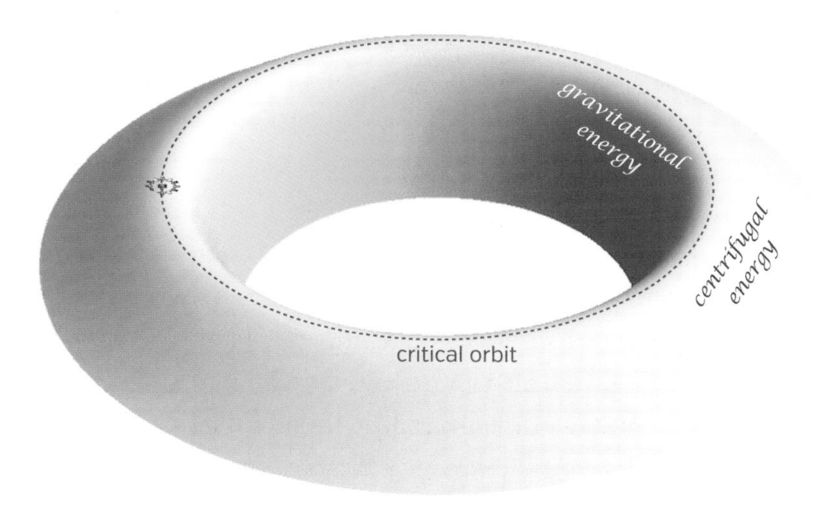

Fig. 27.5. The *Endurance*'s critical orbit on the rim of the volcano, with centrifugal energy and force dominating outside the rim and gravitational energy and force dominating inside. *[Image of the* Endurance *is from* Interstellar.*]*

controlled by the rising centrifugal energy, which has come to dominate over the gravitational energy. On the inside of the volcano, near Gargantua's horizon, the gravitational energy has grown hugely negative and overwhelms the centrifugal energy, so the surface plunges downward (Figure 27.5). The critical orbit is on the volcano's rim.

The Critical Orbit: Balance of Centrifugal and Gravitational Forces

Upon reaching the volcano's rim, the *Endurance,* ideally, would travel around and around it, at constant speed. Because it moves neither inward nor outward, the inward pull of gravity on the rim must precisely be counterbalanced by the outward centrifugal force that arises from the ship's fast circumferential motion.

This indeed is the case, as shown in Figure 27.6—an analog of the force balance plot for Miller's planet (Figure 17.2). At the *Endurance*'s critical orbit, the red curve (the inward gravitational pull on the *Endur-*

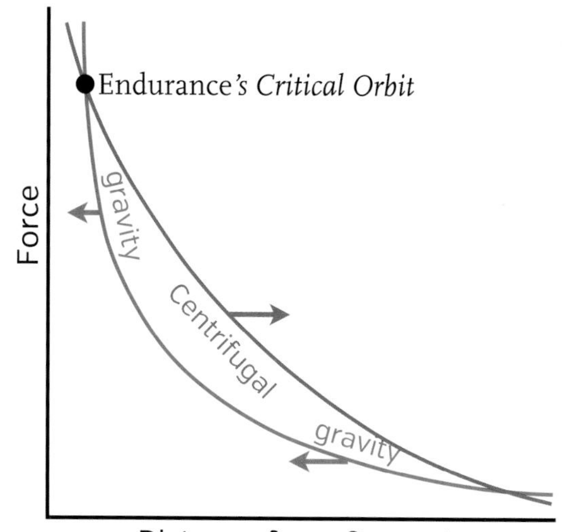

Fig. 27.6. The gravitational and centrifugal forces acting on the *Endurance*, and how they change with changing distance from Gargantua.

ance) and the blue curve (the outward centrifugal force) cross, so the two forces are in balance.

However, the balance is unstable, as our volcano-rim analogy suggests.[2] If the *Endurance* is randomly pushed inward just a bit, then gravity overwhelms the centrifugal force (the red curve rises above the blue curve), so the *Endurance* is pulled on inward toward Gargantua's horizon. If the *Endurance* is pushed outward just a bit, then the centrifugal force wins the battle with gravity (the blue curve is above the red curve), so the *Endurance* is pushed on outward, escaping Gargantua's tight grip.

By contrast (as we saw in Chapter 17), on the orbit of Miller's planet, the balance between the gravitational and centrifugal forces is stable.

Disaster on the Rim: Ejection of TARS and Cooper

In my science interpretation of the movie, the volcano's rim is very narrow, so the critical orbit on the rim is exceedingly unstable. Tiny errors in navigation will send the *Endurance* careening down toward Gargantua (down into the volcano) or away from Gargantua (down toward the moat).

Errors are inevitable, so the *Endurance*'s course must be corrected, continually, by a well-designed feedback system, like an automobile's cruise control but much better.

In my interpretation, the feedback system is not quite good enough and the *Endurance* winds up dangerously far down the inside lip of the volcano. The *Endurance* must use all the thrust at its disposal to climb back up to the critical orbit.

But this is too subtle and technical for action-packed scenes and a hugely diverse audience, so Christopher Nolan chose a simpler, more in-your-face approach. No mention of instability. No mention of feedback. The *Endurance* simply plunges too close to Gargantua, and Cooper responds with all the thrust he can muster to climb back out and escape Gargantua's grip.

The result is the same: lander 1, piloted by TARS, and Ranger 2,

2 The agreement between our volcano-rim analogy and these force arguments is due to a key fact: The net force (gravitational plus centrifugal) on the *Endurance* is proportional to the slope of the energy surface (Figures 27.3 and 27.5). Can you figure out why?

piloted by Cooper, fire their rockets while attached to the *Endurance*, pushing the *Endurance* back out of Gargantua's gravitational grip. Then, to get the last possible kick, explosive bolts blow the *Endurance* apart from lander 1 and Ranger 2. The lander and Ranger go plunging downward toward Gargantua, carrying TARS and Cooper with them, and the *Endurance* is saved (Figures 27.7 and 27.8).

In the movie, there is a tragic, parting conversation between Brand and Cooper. Brand doesn't understand why Cooper and TARS must accompany the lander and Ranger into the black hole. Cooper gives

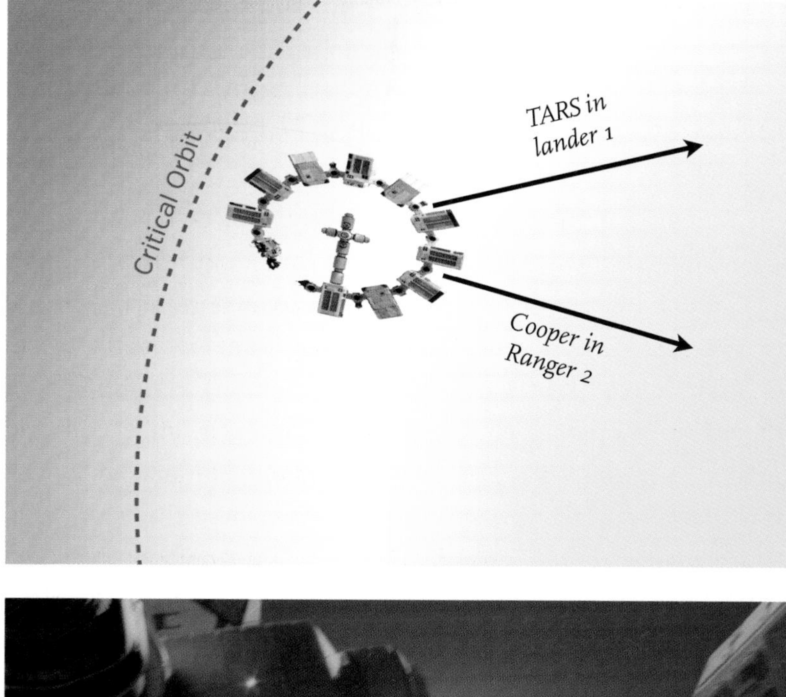

Fig. 27.7. The *Endurance* is thrown back up to the critical orbit by firing of rockets, followed by ejection of lander 1 and Ranger 2. *[Image of the* Endurance *is from* Interstellar.*]*

Fig. 27.8. Ranger 2 descending toward Gargantua, as seen by Brand in the *Endurance*, with portions of two *Endurance* modules in the foreground. The Ranger is the faintly seen object in the picture's lower center, surrounded by Gargantua's accretion disk. *[From* Interstellar, *used courtesy of Warner Bros. Entertainment Inc.]*

her a rather lame though poetic excuse: "Newton's third law. The only way humans have ever figured out for getting somewhere is to leave something behind."

This surely is true. But the additional thrust on the *Endurance*, from Cooper and TARS accompanying the lander and Ranger into the hole, is awfully small. The greater truth, of course, is that Cooper *wants* to go into Gargantua. He hopes that he and TARS can learn the quantum gravity laws from a singularity inside Gargantua, and somehow transmit them back to Earth. It is his last, desperate hope for saving all of humanity.

The *Endurance*'s Launch Toward Edmunds' Planet

The critical orbit is an ideal spot for Brand and the robot CASE to launch the *Endurance* in any desired direction, in particular, toward Edmunds' planet.

How do they control their launch direction? Because the critical orbit is so unstable, a small rocket blast is sufficient to send the *Endurance* off it. And if the blast is ignited at precisely the right location along the critical orbit and has precisely the right strength, it will send the *Endurance* in precisely the desired direction (Figure 27.9).

Actually, Figure 27.9 may leave you unconvinced that Brand and CASE can launch in any direction they wish. That's because it doesn't

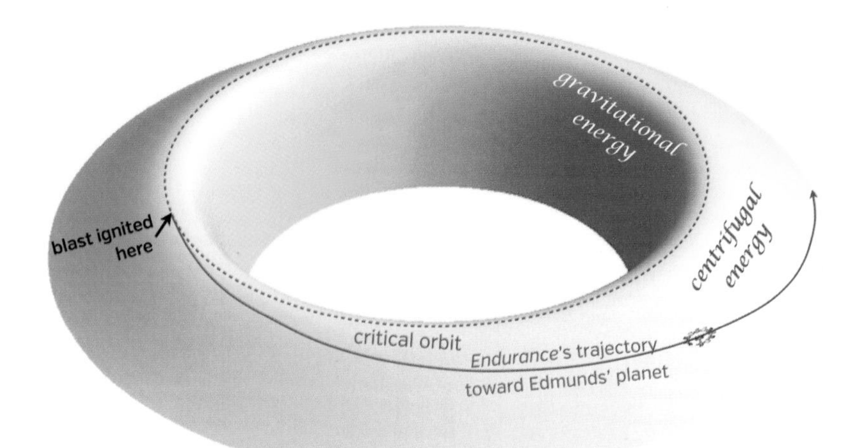

Fig. 27.9. The *Endurance*'s trajectory off the critical orbit, toward Edmunds' planet. *[Image of the* Endurance *is from* Interstellar.*]*

capture the critical orbit's three-dimensional structure. For that, see Figure 27.10.

This convoluted critical orbit is a close analog of the trajectories of temporarily trapped light rays inside Gargantua's shell of fire (Figures 6.5 and 8.2). Like those light rays, the *Endurance* is temporarily trapped when on its critical orbit. Unlike the light rays, the *Endurance* has a control system and rockets, so its launch off the critical orbit is in Brand's and CASE's hands. And because of the orbit's convoluted three-dimensional structure, the launch can be in any direction they wish.

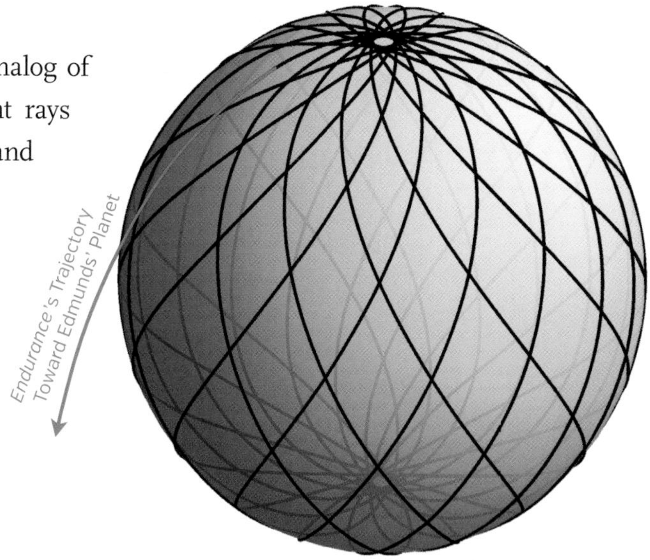

But their launch leaves behind Cooper and TARS, plunging through Gargantua's horizon. Plunging toward Gargantua's singularities.

Fig. 27.10. A three-dimensional picture of the *Endurance*'s critical orbit and its launch toward Edmunds' planet. The critical orbit wraps around a sphere that surrounds Gargantua.

28

Into Gargantua

Some Personal History

Ⓣ

In 1985, when Carl Sagan wanted to send his heroine, Eleanore Arroway (Jodie Foster), through a black hole to the star Vega, I told him *NO!* Inside a black hole she will die. The singularity in the hole's core will tear her apart, chaotically and painfully. I suggested he send Dr. Arroway through a wormhole instead (Chapter 14).

In 2013, I encouraged Christopher Nolan to send Cooper into the black hole Gargantua.

So what happened in the quarter century between 1985 and 2013? Why did my attitude toward falling into a black hole change so dramatically?

In 1985, we physicists thought the cores of all black holes were inhabited by chaotic, destructive BKL singularities, and everything that entered a black hole would be destroyed by the singularity's stretch and squeeze (Chapter 26). That was our highly educated guess. We were wrong.

In the intervening quarter century, two additional singularities were discovered, mathematically, inside black holes: *gentle* singularities, to the extent that any singularity can be gentle (Chapter 26). Gentle enough that Cooper, falling into one, might possibly survive. I'm dubious of survival, but we can't be sure. So I now think it respectable, in science fiction, to posit survival.

Also in the intervening quarter century, we have learned that our universe is probably a brane in a higher-dimensional bulk (Chapter 21). So it's respectable, I think, to posit living beings that inhabit the bulk—a very advanced civilization of bulk beings—who might save Cooper from the singularity at the last moment. That's what Christopher Nolan chose.

Through the Event Horizon

Ⓣ

In *Interstellar,* when Ranger 2 piloted by Cooper (and lander 1, piloted by TARS) eject from the *Endurance,* they spiral down toward Gargantua's event horizon and then through it. What do Einstein's relativistic laws say about this downward spiral?

According to those laws, and hence my interpretation of the movie, Brand, watching from the *Endurance,* can never see the Ranger penetrate the horizon. No signal Cooper tries to send her from inside the horizon can ever get out. The flow of time inside the horizon is downward, and that downward time flow drags Cooper and all signals he sends downward with itself, away from the horizon. See Chapter 5.

So what *does* Brand see (if she and CASE can stabilize the *Endurance* long enough for her to watch)? Because the *Endurance* and the Ranger are both deep in the cylindrical part of Gargantua's warped space (Figure 28.1), they are both dragged circumferentially by Gargantua's whirling space with almost the same angular velocity (the same orbital period). So as seen by Brand, in her orbiting reference frame, the Ranger drops away from the *Endurance* almost straight downward toward the horizon (Figure 28.1). That's what the movie depicts.

As Brand watches the Ranger approach the horizon, she must see time on the Ranger slow and then

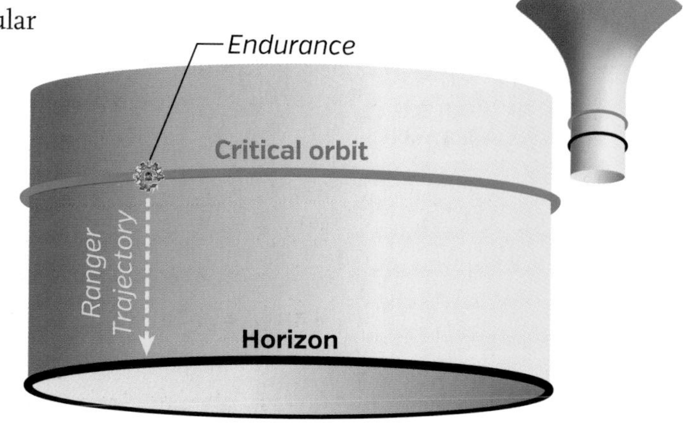

Fig. 28.1. The Ranger's trajectory through Gargantua's warped space, as seen in the *Endurance*'s orbiting reference frame. The *Endurance* is drawn far larger than it should be, so you can see it. *Inset*: A larger portion of Gargantua's warped space. *[Image of the* Endurance *is from* Interstellar.*]*

freeze relative to her time, Einstein's laws say. This has several consequences: She sees the Ranger slow its downward motion and then freeze just above the horizon. She sees light from the Ranger shift to longer and longer wavelengths (lower and lower frequencies, becoming redder and redder), until the Ranger turns completely black and unobservable. And bits of information that Cooper transmits to Brand one second apart as measured by his time on the Ranger arrive with larger and larger time separations as measured by Brand. After a few hours Brand receives the last bit that she will ever receive from Cooper, the last bit that Cooper emitted before piercing the horizon.

Cooper, by contrast, continues receiving signals from Brand even after he crosses the horizon. Brand's signals have no trouble entering Gargantua and reaching Cooper. Cooper's signals can't get out to Brand. Einstein's laws are unequivocal. This is how it must be.

Moreover, those laws tell us that Cooper sees nothing special as he crosses the horizon. He can't know, at least not with any ease, which bit that he transmits is the last one Brand will receive. He can't tell, by looking around himself, precisely where the horizon is. The horizon is no more distinguishable to him than the Earth's equator is to you as you cross it in a ship.

These seemingly contradictory observations by Brand and Cooper are a result of two things: The warping of time, and the finite travel time for the light and information that they send to each other. When I think carefully about *both* of these things, I don't see any contradiction at all.

Sandwiched Between Singularities

(EG)

As the Ranger carries Cooper deeper and deeper into the bowels of Gargantua, he continues to see the universe above himself. Chasing the light that brings him that image is an infalling singularity. The singularity is weak at first, but it grows stronger rapidly, as more and more stuff falls into Gargantua and piles up in a thin sheet (Chapter 27). Einstein's laws dictate this.

Below the Ranger is an outflying singularity, created by stuff that fell into the black hole long ago and was backscattered upward toward the Ranger (Chapter 27).

The Ranger is sandwiched between the two singularities (Figure 28.2). Inevitably, it will be hit by one or the other.

When I explained the two singularities to Chris, he immediately knew which one should hit the Ranger. The outflying singularity. Why? Because Chris had already adopted, for *Interstellar,* a variant of the laws of physics that prevents physical objects from ever traveling backward in time (Chapter 30). The infalling singularity is produced by stuff that falls into Gargantua long after Cooper falls in (long after, as measured by the external universe's time; Earth's time). If Cooper is hit by that singularity and survives, the universe's far future will be in his past. He will be so far in *our* future that, even with the help of bulk beings, he won't be able to return to the solar system until billions of years after he left, if ever. That would prevent him from ever reuniting with his daughter, Murph.

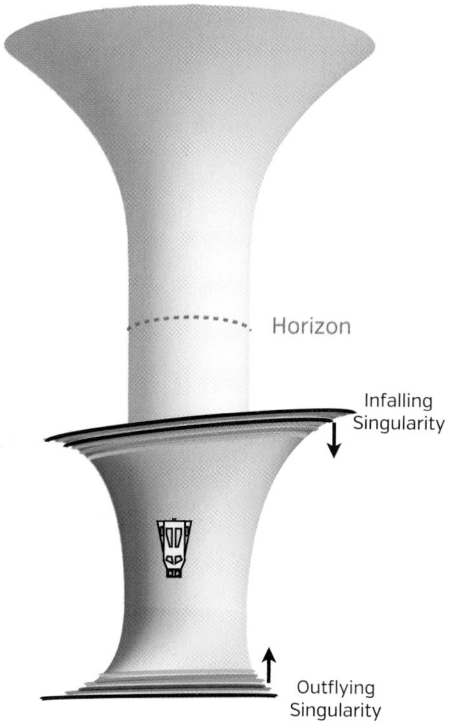

Fig. 28.2. An icon representing the Ranger sandwiched between Gargantua's infalling and outflying singularities. The Ranger is drawn far larger than it should be, so you can see it.

So Chris firmly chose Cooper to be hit by the outflying singularity, not the infalling one—hit by the singularity arising from stuff that fell into Gargantua *before* the Ranger, not after it.

Chris's choice, though, presents a bit of a problem for my scientist's interpretation of the movie. But not a problem so severe as backward time travel. If the Ranger falls directly into Gargantua from the critical orbit, then its infall is slow enough that the infalling singularity will catch up to it and hit it. For the Ranger to hit the outflying singularity instead, as Chris wants, the Ranger must nearly outrun the infalling singularity, which is descending at the speed of light. The Ranger can do so, if it is given a large, inward kick. How? The usual: by a slingshot around a suitable intermediate-mass black hole soon after leaving the *Endurance.*

What Does Cooper See Inside Gargantua?

Ⓢ

Looking up as he falls inward, Cooper sees the external universe. Because his infall has been sped up, he sees time in the external universe flow at roughly the same rate as his own time,[1] and he sees the image of the external universe reduced in size,[2] from about half of the sky to roughly a quarter.

When I was first shown the movie's depiction of this, I was pleased to discover that Paul Franklin's team got it right, and also got right something I had missed: In the movie, the image of the universe above is surrounded by Gargantua's accretion disk (Figure 28.3). Can you explain why this must be so?

Cooper sees all this above him, but he doesn't see the infalling singularity. It is moving downward toward him at the speed of light, chasing but not catching the light rays that bring him images of the disk and universe above.

Because we are rather ignorant of what goes on inside black holes, I told Chris and Paul that I'd be comfortable if they used their imaginations in depicting what Cooper sees coming up at him from below, as he falls. I made only one request: "Please don't depict Satan and the fires of Hades inside the black hole like the Disney Studios did in their *Black Hole* movie." Chris and Paul chuckled. They weren't tempted in the least.

When I saw what they did depict, it made great sense. Looking

Fig. 28.3. The universe above, surrounded by the accretion disk, as seen by Cooper inside Gargantua, looking upward across his Ranger's fuselage. Gargantua's shadow is the black region on the left. *[From* Interstellar, *used courtesy of Warner Bros. Entertainment Inc.]*

1 In technical language, signals from above are Doppler shifted to the red by his high speed, which compensates the blue shift produced by the hole's gravitational pull, so colors look fairly normal.

2 Due to aberration of the starlight.

downward, Cooper should see light from objects that fell into Gargantua before him and are still falling inward. Those objects need not emit light themselves. He can see them in reflected light from the accretion disk above, just as we see the Moon in reflected sunlight. I expect those objects to be mostly interstellar dust, and this could explain the fog he encounters in the movie as he falls.

Cooper can also overtake stuff that's infalling more slowly than he. This may explain the white flakes that hit and bounce off his Ranger in the movie.

Rescued by the Tesseract

(S)

In my science interpretation, as the Ranger nears the outflying singularity, it encounters mounting tidal forces. Cooper ejects just in the nick of time. Tidal forces tear the Ranger apart. Visually, it splits in two.

At the singularity's edge the tesseract awaits Cooper—placed there, presumably, by bulk beings (Figure 28.4).

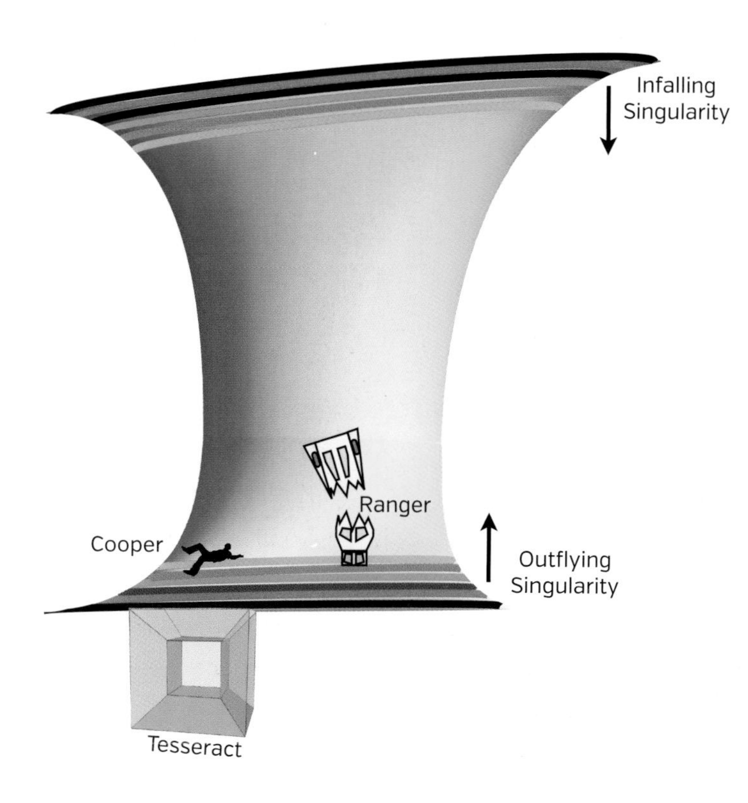

Fig. 28.4. An icon representing Cooper about to be scooped up by the tesseract at the edge of the singularity. The Ranger icon and Cooper icon are drawn far, far larger than they should be, so you can see them, and are drawn two-dimensional, since one space dimension is suppressed from this diagram.

The Tesseract

In *Interstellar,* the entrance to the tesseract is a white checkerboard pattern. Each white square is the end of a beam. Cooper, entering the tesseract, falls down a channel between beams, dazed and confused, lashing out at what appear to be bricks along the channel wall, but turn out to be books. The channel leads to a large chamber, where he floats and struggles, gradually getting oriented.

The chamber is Christopher Nolan's unique take on one three-dimensional face of the four-dimensional tesseract, enhanced by Paul Franklin and his visual-effects team. The chamber and its environs are remarkably complex. Seeing them for the first time, I felt as disoriented as Cooper, even though I know what a tesseract is. Chris and Paul had enriched the tesseract so greatly that I only fully understood after talking with them.

Here's what I know—and what I learned, filtered through my physicist's eyes. I begin with the standard, simple tesseract, and then I build up to Chris's complexified tesseract.

From Point to Line to Square to Cube to Tesseract

Ⓣ

A standard tesseract is a hypercube, a cube in four space dimensions. In Figures 29.1 and 29.2 I walk you through what this means.

If we take a point (top of Figure 29.1) and move it in one dimension, we get a line. The line has two faces (ends); they are points. The line has one dimension (it extends along one dimension); its faces have one less dimension: zero.

If we take a line and move it in a dimension perpendicular to itself (middle of Figure 29.1), we get a square. The square has four faces; they are lines. The square has two dimensions; its faces have one less dimension: one.

If we take a square and move it in a dimension perpendicular to itself (bottom of Figure 29.1), we get a cube. The cube has six faces; they are squares. The cube has three dimensions; its faces have one less dimension: two.

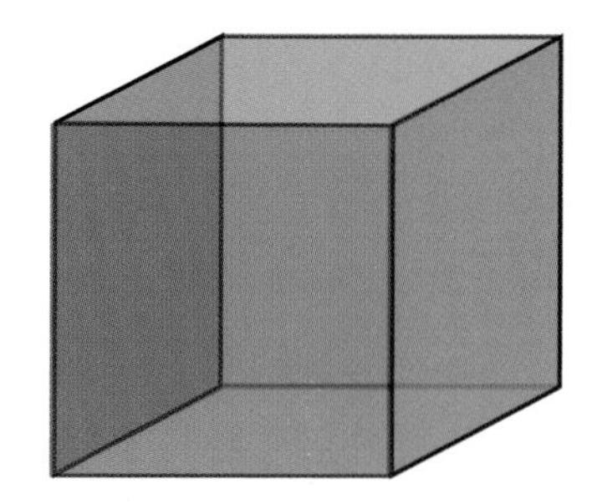

Fig. 29.1. From point to line to square to cube.

The next step should be obvious, but to visualize it, I need to redraw the cube as you would see it if you were up close to one of the orange faces (top of Figure 29.2). Here the original square (the small, dark orange one), when moved toward you to form the cube, appears to enlarge to become the cube's front face, the outer square.

If we take a cube and move it in a dimension perpendicular to itself (bottom of Figure 29.2), we get a tesseract. The picture of the tesseract is analogous to the one above it, of the cube: It looks like two cubes, inside each other. The inner cube has expanded outward, in the picture, to sweep out the four-dimensional volume of the tesseract. The tesseract has eight faces; they are cubes. (Can you identify and count them?) The tesseract has four space dimensions; its faces have one less dimension: three. The tesseract and its faces share one time dimension, not shown in the picture.

The chamber Cooper enters in the film is one of the tesseract's eight cubical faces, though, as I said earlier, modified in a clever, complex way by Chris and Paul. Before explaining their clever modifications, I use the standard, simple tesseract to describe my interpretation of the movie's early tesseract scenes.

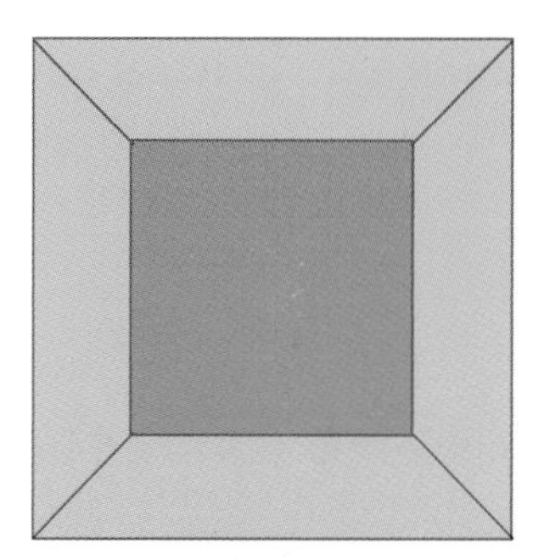

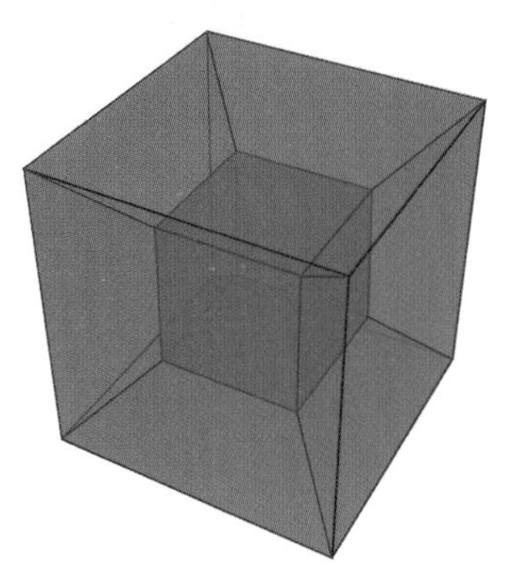

Fig. 29.2. From cube to tesseract.

Cooper Transported in the Tesseract

Because Cooper is made of atoms held together by electric and nuclear forces, all of which can exist only in three space dimensions and one time, he is confined to reside in one of the tesseract's three-space-dimensional faces (cubes). He can't experience the tesseract's fourth spatial dimension. Figure 29.3 shows him floating in the tesseract's front face, whose edges I delineated by purple lines.

In my interpretation of the movie, the tesseract ascends from the singularity into the bulk. Being an object with the same number of space dimensions as the bulk (four), it happily inhabits the bulk. And it transports three-dimensional Cooper, lodged in its three-dimensional face, through the bulk.

Now, recall that the distance from Gargantua to Earth is about 10 billion light-years as measured in our brane (our universe, with its three space dimensions). However, as measured in the bulk, that distance is only about 1 AU (the distance from the Sun to the Earth); see Figure 23.7. So, traveling with whatever propulsion system the bulk beings provided, the tesseract, in my interpretation, can quickly carry Cooper across our universe, via the bulk, to Earth.

Fig. 29.3. A Cooper icon in a three-dimensional face of the tesseract.

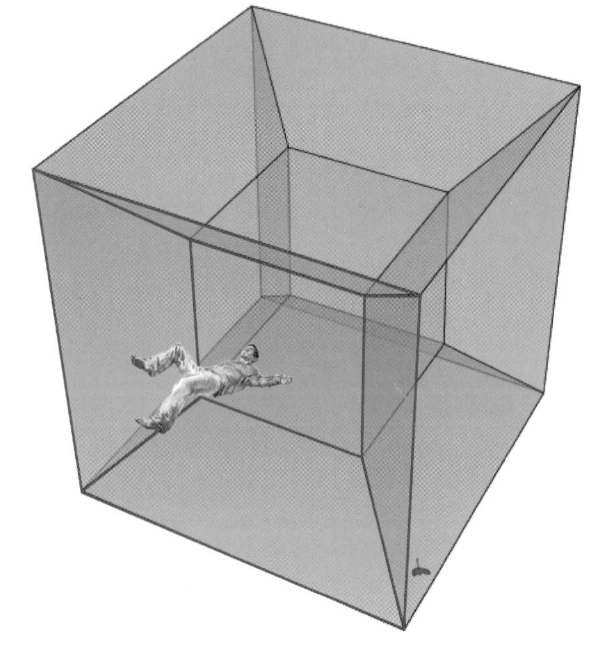

Figure 29.4 is a snapshot from that trip. One spatial dimension is suppressed from the snapshot, so the tesseract is a three-dimensional cube in a three-dimensional bulk, and Cooper has become a two-dimensional icon of a man, in a two-dimensional face of the cube, traveling parallel to our two-dimensional universe (brane).

To match what is shown in the movie, I imagine this trip is very quick, just a few minutes, while Cooper is still dazed and falling. As he comes to rest, floating in the large chamber, the tesseract docks beside Murph's bedroom.

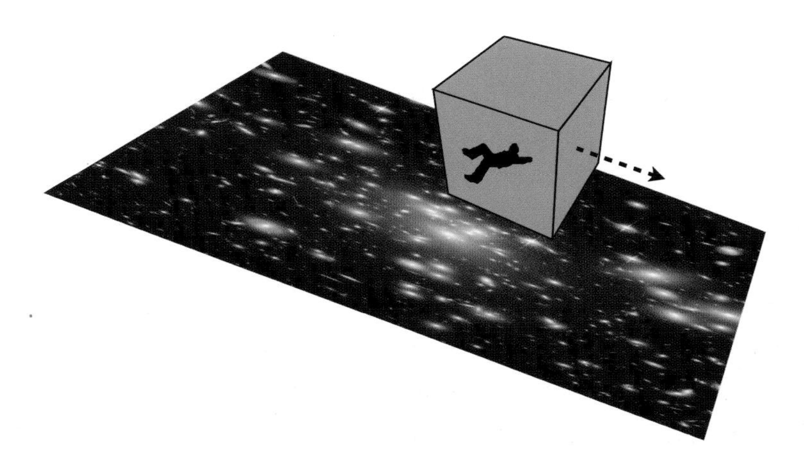

Fig. 29.4. The Cooper icon transported through the bulk, above our brane, riding in a face of the tesseract. One space dimension is removed from this picture.

Docking: The View into Murph's Bedroom

How does this docking work? In my interpretation, arriving in the bulk near Earth the tesseract must penetrate the 3-centimeter-thick AdS layer that encases our brane (Chapter 23) in order to reach Murph's bedroom. Presumably the bulk beings who built the tesseract equipped it with technology to push the AdS layer to the side, clearing the way for its descent.

Figure 29.5 shows the tesseract, after the clearing, docked alongside Murph's bedroom in Cooper's farmhouse. Again, one spatial dimension is suppressed, so the tesseract is depicted as a three-dimensional cube and the farmhouse and bedroom and Murph are two dimensional, as, of course, is Cooper.

The back face of the tesseract coincides with Murph's bedroom. I'll explain that more carefully. The back face is a three-dimensional cross section of the tesseract that resides in Murph's bedroom in the same sense as the circular cross section of a sphere resides in a two-dimensional brane in Figure 22.2, and a spherical cross section of a hypersphere resides in a three-dimensional brane in Figure 22.3. So everything in Murph's bedroom, including Murph herself, is also inside the tesseract's back face.

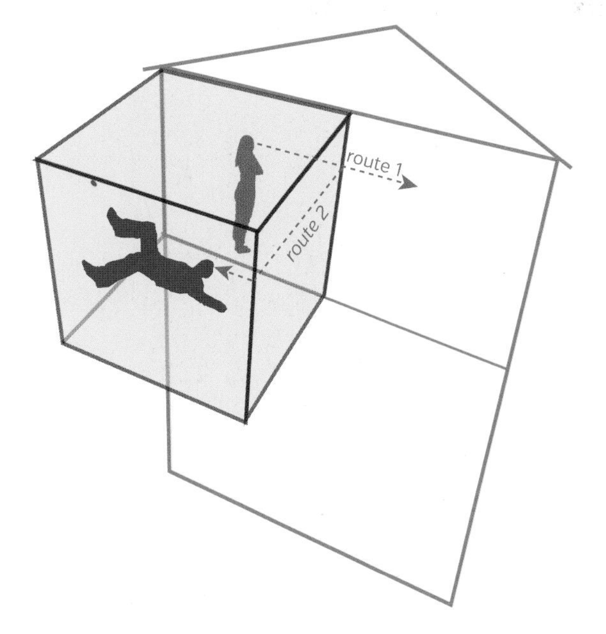

Fig. 29.5. The tesseract docked alongside Murph's bedroom.

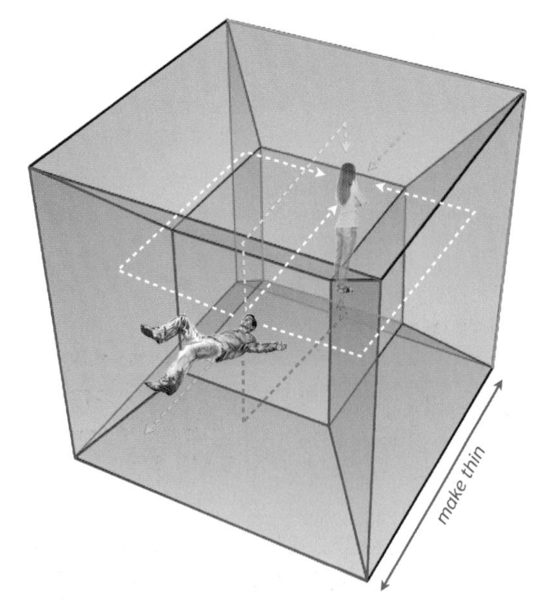

Fig. 29.6. The Cooper icon can see into Murph's bedroom (*orange edges*) by looking through each of the six walls of his face of the tesseract (*purple edges*). Here he sees an icon of Murph herself.

When a light ray traveling out from Murph reaches the common edge of Murph's bedroom and the tesseract, it has two places to go: The ray can stay in our brane, traveling along route 1 of Figure 29.5 out an open door or into a wall where it is absorbed. Or the ray can stay in the tesseract, traveling along route 2 into and through the next tesseract face, and then onward to Cooper's eyes. Some of the ray's photons go along route 1; others go along route 2, bringing Cooper an image of Murph.

Now look at Figure 29.6, in which I restore the suppressed dimension. When Cooper looks through the right wall of his chamber, he sees into Murph's bedroom through its right wall (right white light ray). Looking through the left wall of his chamber, Cooper sees into Murph's bedroom through its left wall (left white light ray). Looking through his back wall, he sees into the bedroom through its back wall. Looking through his front wall (orange light ray), he sees into the bedroom through its front wall (though this is not obvious in Figure 29.6; can you explain why it is true?). Looking along the yellow ray, he sees down through her ceiling. Looking along the red ray, he sees up through her floor. To Cooper, as he changes his gaze from one direction to another to another, it seems like he is orbiting Murph's bedroom. (This is how Chris described it when he first showed me his complexified tesseract.)

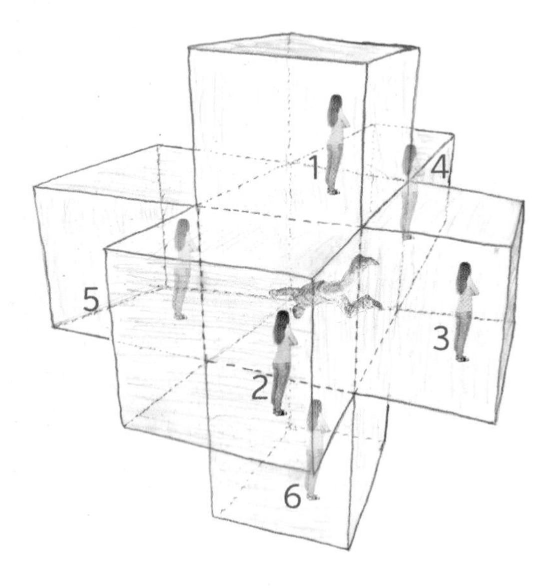

Fig. 29.7. The six views of Murph's bedroom seen by the Cooper icon from his tesseract face. *[My own hand sketch.]*

In Figure 29.6, all six light rays have to pass through intermediate cubes (tesseract faces) before reaching Murph's bedroom. In the movie they don't travel any noticeable distance from chamber to bedroom, so Chris and Paul must have shrunk the tesseract in one dimension; see the gray arrow and notation "make thin" in Figure 29.6.

After that shrinkage, every face of Cooper's chamber looks directly and immediately into one of the faces (wall or floor or ceiling) of Murph's bedroom with no intervening space, so to Cooper the situation looks like Figure 29.7. He sees six bedrooms, one bordering each face of his chamber but all identical except for his viewing direction.[1] In fact they are all identical. There is only one bedroom, although to Cooper there appear to be six.

Nolan's Complexified Tesseract

Figure 29.8 is a still, showing Cooper floating in his chamber inside the tesseract. It looks very different from Figure 29.7 because of the complex and rich modifications that Chris conceived, and Paul and his team implemented.

Fig. 29.8. Cooper floating in Nolan's complexified tesseract. *[From* Interstellar, *used courtesy of Warner Bros. Entertainment Inc.]*

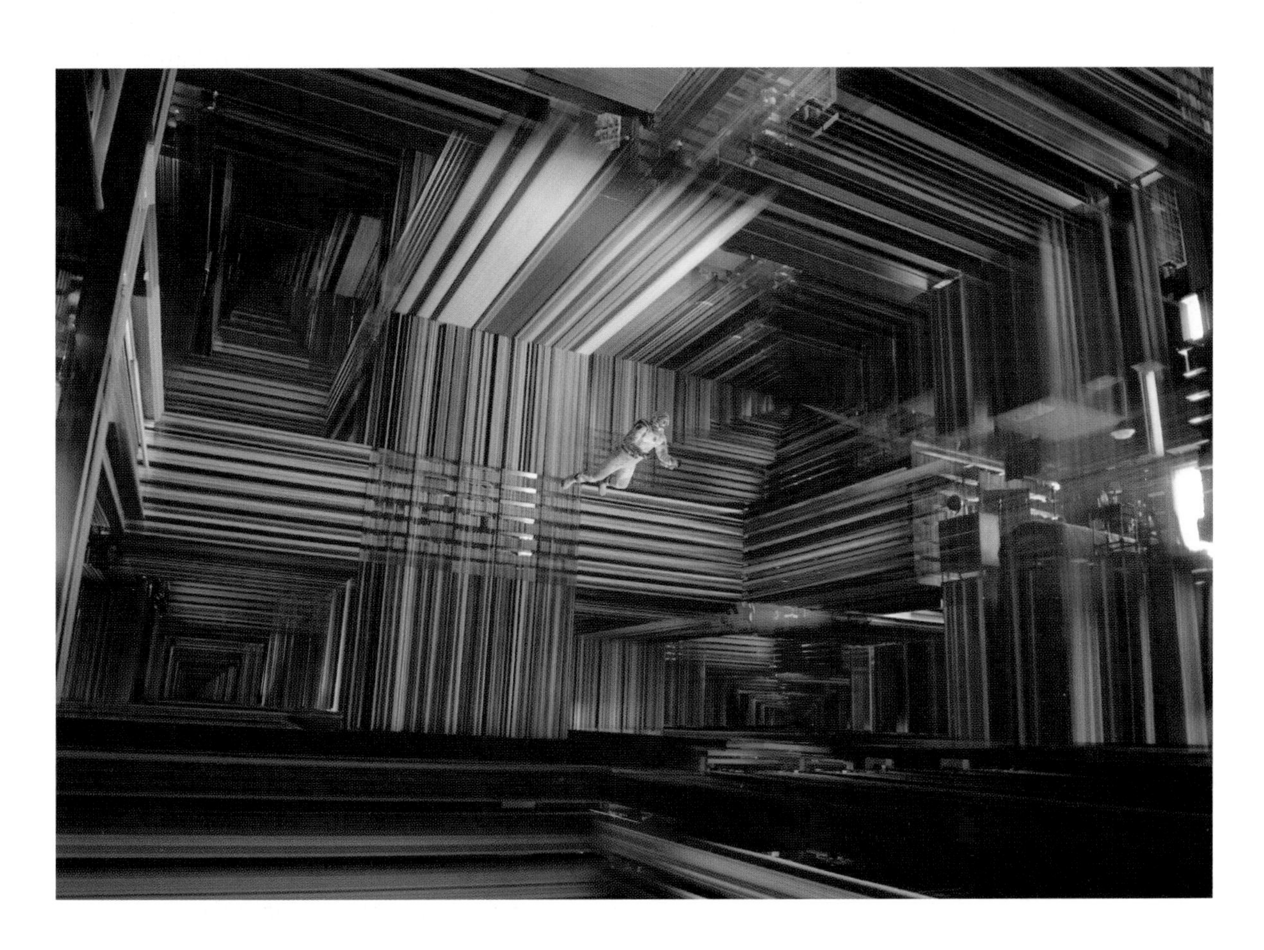

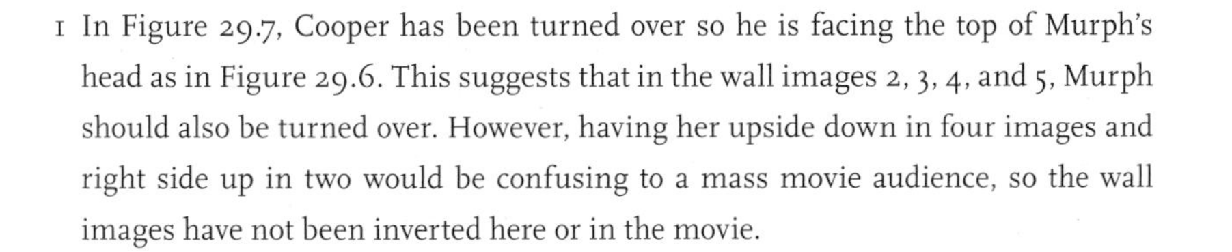

1 In Figure 29.7, Cooper has been turned over so he is facing the top of Murph's head as in Figure 29.6. This suggests that in the wall images 2, 3, 4, and 5, Murph should also be turned over. However, having her upside down in four images and right side up in two would be confusing to a mass movie audience, so the wall images have not been inverted here or in the movie.

Fig. 29.9. The size of Cooper's chamber enlarged threefold so the six bedrooms occupy the centers of his chamber's faces. *[My own hand sketch.]*

The first thing I noticed when I saw Chris's complexified tesseract was the threefold enlargement of Cooper's chamber, so the bedroom attached to each chamber face covers only a third of the face. I depict this in Figure 29.9 with all the other tesseract complexities removed and the chamber's back three faces hidden from view.[2]

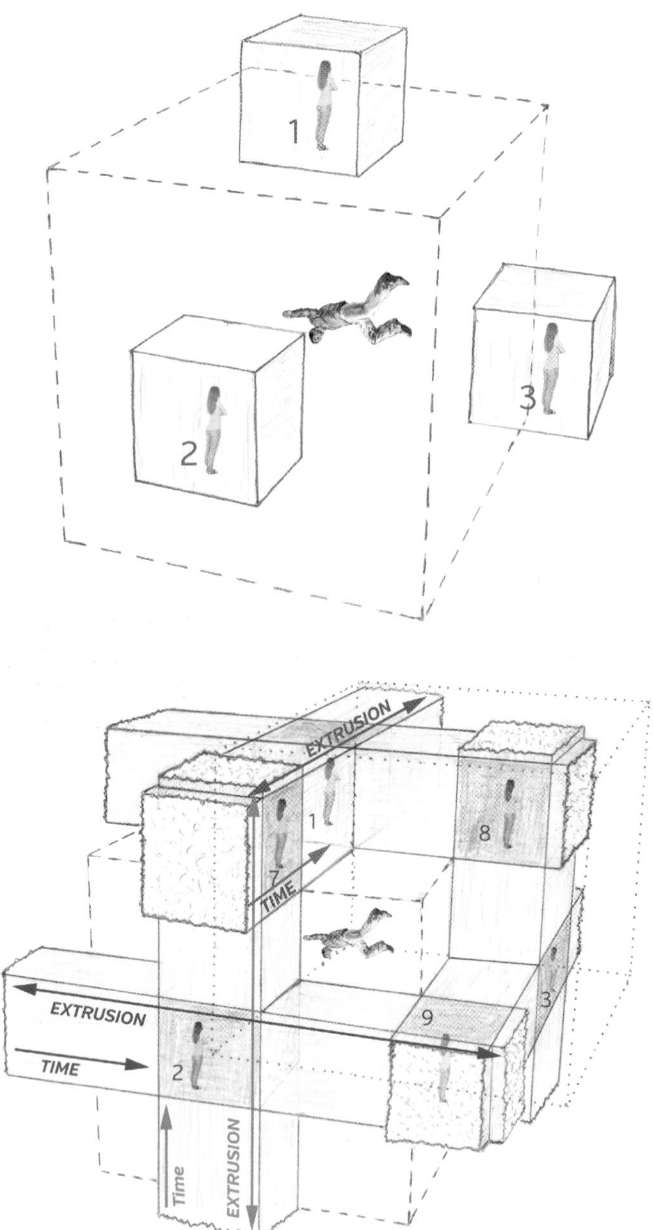

The next thing I noticed were two extrusions extending out of each bedroom along the two directions transverse to Cooper's chamber (Figures 29.10 and 29.11). As Chris and Paul explained it to me, wherever these extrusions intersect there is a bedroom; for example, bedrooms 7, 8, and 9 as well as the original 1–6.

The extrusions extend indefinitely, creating at their intersections a seemingly infinite lattice of bedrooms and of chambers[3] like Cooper's [dashed edges in Fig. 29.10.]. For example, the labeled faces of bedrooms 7, 8, and 9 face into a chamber whose edges are indicated with dots; the back-left-bottom corner of that chamber overlaps the front-right-top corner of Cooper's chamber.

TARS gives us a clue to the meaning of the extrusions and the latticework of bedrooms and chambers when he tells Cooper, "You've seen that time is represented here as a physical dimension."

Chris and Paul elaborated on that

Fig. 29.10. Extrusions extend out of all the bedrooms, and time flows along them. *[My own hand sketch.]*

2 In the movie Murph's bedroom is not a cube; its length, width, and height are 20, 15, and 10 feet, and Cooper's chamber is three times larger in each dimension: 60, 45, and 30 feet. For simplicity, I idealize the bedrooms and chambers as cubes.

3 Chris and Paul call these chambers "voids" because they are regions through which no extrusions pass.

Fig. 29.11. The lattice of extrusions, drawn by Christopher Nolan in his working notebook when developing the concepts for the complexified tesseract.

clue for me. The bulk beings, they explained, are displaying time for the blue extrusions as flowing along the blue-arrowed direction in Figure 29.10, and for the green extrusions along the green-arrowed direction, and for the brown extrusions along the brown-arrowed direction.

To understand this in greater detail, let's focus momentarily on the single pair of extrusions that intersect in bedroom 2; see Figure 29.12. Cross sections through the room that are vertical in the picture travel rightward with passing time, along the blue time arrow; and as they travel, they create the blue extrusion. Similarly, cross sections that are horizontal travel upward as time passes, along the green time arrow, creating the green extrusion. Where the two sets of cross sections intersect—where the extrusions intersect—there is a bedroom.

The same is true for all other extrusions. At each intersection of two extrusions, the cross sections they carry produce a bedroom.

Because of the cross sections' finite speed, the various bedrooms are out of time synch

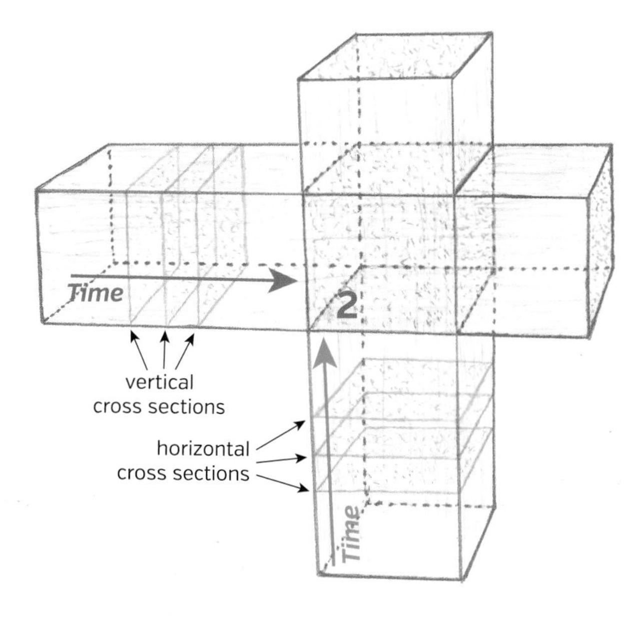

Fig. 29.12. Cross sections of Murph's bedroom travel along two extrusions. Bedroom 2 resides where the two sets of cross sections intersect. *[My own hand sketch.]*

with each other. For example, if it takes one second for cross sections to travel along each extrusion from one bedroom to the next, then all the bedrooms in Figure 29.13 are to the future of image 0 by the number of seconds shown in black. In particular, bedroom 2 is one second ahead of bedroom 0, bedroom 9 is two seconds ahead of bedroom 0, and bedroom 8 is four seconds ahead of bedroom 0. Can you explain why?

In the movie, the time lapse between adjacent bedrooms is closer to a tenth of a second than a full second. By watching adjacent bedrooms carefully as the curtains in Murph's bedroom window blow in the wind, you can estimate the time between bedrooms.

Of course each bedroom in the movie's tesseract is Murph's actual bedroom at a particular moment of time—the time labeled in black in Figure 29.13.

Cooper can move far faster than the flow of time in the bedroom extrusions, so he can easily travel through the tesseract complex to most any bedroom time that he wishes!

To travel most rapidly into the future of Murph-bedroom time, Cooper should move along a diagonal of his chamber in the direction of increasing blue, green, and brown time (rightward, upward, and inward)—that

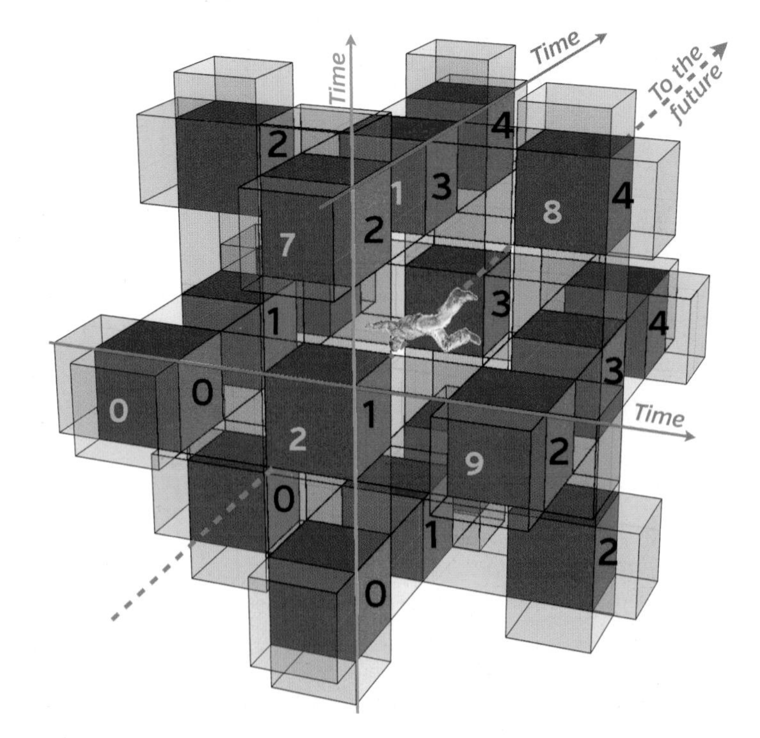

Fig. 29.13. A portion of the lattice of bedrooms created by the intersections of the moving cross sections (the extrusions). The blue numbers identify specific bedrooms—an extension of the numbering system in previous figures. The black number on each bedroom indicates its amount of time *to the future* of bedroom 0. The dashed violet arrow is the direction in which Cooper can move most rapidly into the bedroom's future.

is, along the diagonal dashed violet line in Figure 29.13. Diagonals like this are devoid of extrusions; they are open channels along which Cooper can travel. In the movie we see him traveling along such an open diagonal channel to get from the bedroom time of the early ghostly book falls to the bedroom time of the wristwatch ticking (Figure 29.14).

Is Cooper really traveling forward and backward in time as he moves diagonally up and down through the complex? Forward and backward in the manner that Amelia Brand speculates bulk beings can when she says: "To Them time may be just another physical dimension. To Them the past might be a canyon they can climb into and the future a mountain they can climb up. But to us it's not. Okay?"

What *are* the rules governing time travel in *Interstellar*?

Fig. 29.14. This is what Cooper sees as he travels rapidly into the future of Murph-bedroom time by soaring along a diagonal channel through the tesseract complex. The diagonal channel is in the picture's upper center. *[From* Interstellar*, used courtesy of Warner Bros. Entertainment Inc.]*

30

Messaging the Past

Communicating Rule Sets to a Movie Audience

Ⓣ

Before Christopher Nolan became *Interstellar*'s director and rewrote the screenplay, his brother Jonah taught me about rule sets.

To maintain the desired level of suspense in a science-fiction movie, Jonah said, the audience must be told the rules of the game, the movie's "rule set." What do the laws of physics and the technology of the era allow, and what do they forbid? If the rules are not clear, then many in the audience will expect some miraculous event to save the heroine, out of the blue, and tension will fail to mount as it should.

Of course you can't say to the audience, "Here is the rule set for this movie: . . ." It must be communicated in a subtle and natural way. And Chris is a master of this. He communicates his rule sets though the characters' dialog. Next time you watch *Interstellar* (how can you resist watching it again?), look within the film for his telltale bits of rule-set dialog.

Christopher Nolan's Rule Set for Time Travel

It turns out (see below) that backward time travel is governed by the laws of quantum gravity, which are terra almost incognita, so we physicists don't know for sure what is allowed and what not.

Chris made two specific choices for allowed and forbidden time travel—his rule set:

Rule 1: Physical objects and fields with three space dimensions, such as people and light rays, cannot travel backward in time from one location in our brane to another, nor can information that they carry. The physical laws or the actual warping of spacetime prevent it. This is true whether the objects are forever lodged in our brane or journey through the bulk in a three-dimensional face of a tesseract, from one point in our brane to another. So, in particular, Cooper can never travel to his own past.

Rule 2: Gravitational forces can carry messages into our brane's past.

In the movie, rule 1 generates mounting tension. Murph grows older and older as Cooper lingers near Gargantua. With no possibility to travel backward in time there's a growing danger he'll never return to her.

Rule 2 gives Cooper hope. Hope that he can use gravity to transmit the quantum data backward in time to young Murph, so she can solve the Professor's equation and figure out how to lift humanity off Earth.

How do these rules play out in *Interstellar*?

Messaging Murph

When falling into and through the tesseract, Cooper truly does travel backward relative to our brane's time, from the era when Murph is an old woman to the era when she is ten years old. He does this in the sense that, looking at Murph in the tesseract bedrooms, he sees her ten years old. And he can move forward and backward relative to our brane's time (the bedroom's time) in the sense that he can look at Murph at various bedroom times by choosing which bedroom to look into. This does not violate rule 1 because Cooper has not reentered our brane. He remains outside it, in the tesseract's three-dimensional channel, and he looks into Murph's bedroom via light that travels forward in time from Murph to him.

But just as Cooper can't reenter our brane in Murph's ten-year-old era, so he can't send light to her. That would violate rule 1. The light could bring her information from Cooper's personal past, which is her future; information from the era when she is an old woman—backward-in-time information from one location in our brane to another. So there must be some sort of one-way spacetime barrier between ten-year-old Murph in her bedroom and Cooper in the tesseract, rather like a one-way mirror or a black-hole horizon. Light can travel from Murph to Cooper but not from Cooper to Murph.

In my scientist's interpretation of *Interstellar,* the one-way barrier has a simple origin: Cooper, in the tesseract, is always in ten-year-old Murph's future. Light can travel toward the future from Murph to him. It can't travel to the past from him to Murph.

However, gravity can surmount that one-way barrier, Cooper discovers. Gravitational signals can go backward in time from Cooper to Murph. We first see this when Cooper desperately pushes books out of Murph's bookcase. Figure 30.1 shows a still from that scene of the movie.

To explain this still, I must tell you a bit more about the bedroom extrusions, as Chris and Paul Franklin explained them to me. Let's focus on the front blue extrusion in Figures 29.10 and 29.12, which I reproduce as Figure 30.2 with extraneous stuff removed. Recall that this extrusion is a set of vertical cross sections through Murph's bedroom, traveling forward in bedroom time along the blue direction (rightward).

Each object in the bedroom, for example each book, contributes to the bedroom's extrusion. In fact, the book has its own extrusion, which travels forward in time along the blue-arrow direction as part of the bedroom's larger extrusion. We physicists call a variant of this

Fig. 30.1. Cooper pushes on the world tube of a book with his right hand. *[From* Interstellar, *used courtesy of Warner Bros. Entertainment Inc.]*

extrusion the book's "world tube." And we call the extrusion of each particle of matter in the book the particle's "world line." So the book's world tube is a bundle of world lines of all the particles that make up the book. Chris and Paul also use this language. The thin lines that you see in the movie, running along the extrusions, are world lines of particles of matter in Murph's bedroom.

In Figure 30.1, Cooper slams his fist on the book's world tube over and over again, creating a gravitational force, which travels backward

Fig. 30.2. The world tube of a book, within an extrusion of Murph's bedroom. The book and its world tube are drawn much larger than they actually are. *[My own hand sketch.]*

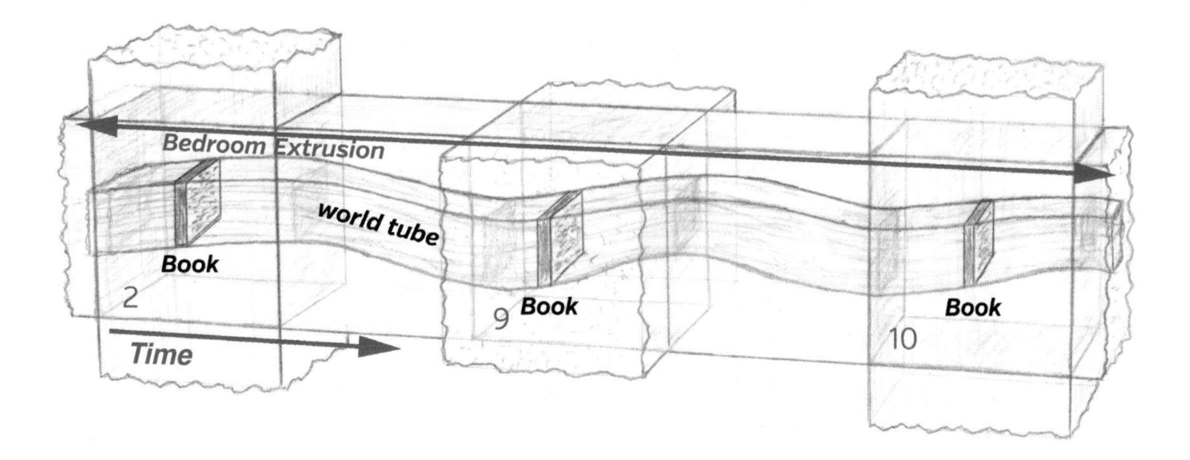

in time to the moment in Murph's bedroom that he is seeing and then pushes on the book's world tube. The book's tube responds by moving. The tube's motion appears to Cooper as an instantaneous response to his pushes. And the motion becomes a wave traveling leftward down the tube (Figure 30.2).[1] When the motion gets strong enough, the book falls out of the bookcase.

By the time Cooper has received the quantum data from TARS, he has mastered this means of communication. In the movie we see him pushing with his finger on the world tube of a watch's second hand. His pushes produce a backward-in-time gravitational force, which makes the second-hand twitch in a Morse-encoded pattern that carries the quantum data. The tesseract stores the twitching pattern in the bulk so it repeats over and over again. When forty-year-old Murph returns to her bedroom three decades later, she finds the second hand still twitching, repeating over and over again the encoded quantum data that Cooper has struggled so hard to send her.

How does the backward-in-time gravitational force work? I'll describe my physicist's interpretation after I tell you what I know, or think I know, about backward time travel.

Time Travel Without a Bulk: What I Think I Know

(EG)

In 1987, triggered by Carl Sagan (Chapter 14), I realized something amazing about wormholes. If wormholes are allowed by the laws of physics, then Einstein's relativistic laws permit transforming them into time machines. The nicest example of this was discovered a year later by my close friends Valery Frolov and Igor Novikov, in Moscow, Russia. Their example, Figure 30.3, shows that a wormhole's transformation into a time machine might occur naturally, without the aid of intelligent beings.

In Figure 30.3, the bottom mouth of the wormhole is in orbit around a black hole and the upper mouth is far from the black hole.

1 Why leftward? So the tube is always at the same transverse position at any specific moment of bedroom time. Think about it.

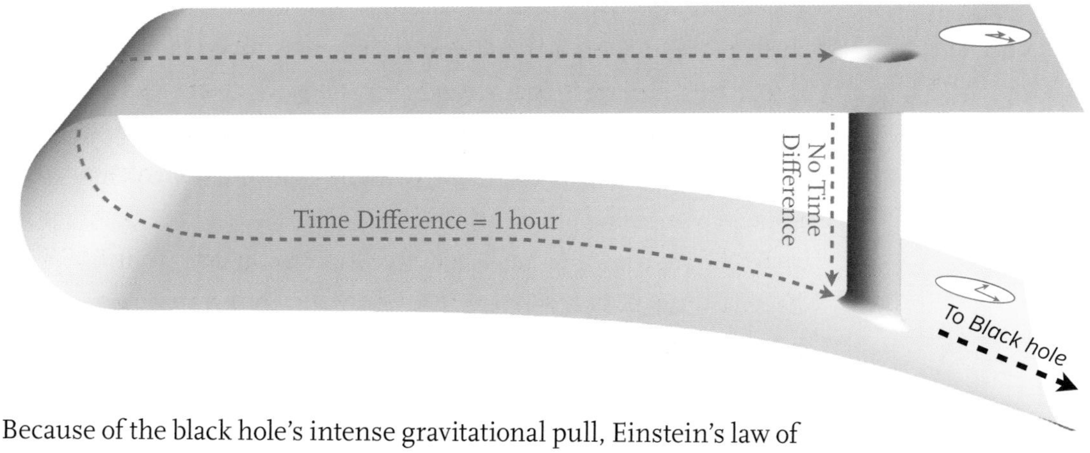

Because of the black hole's intense gravitational pull, Einstein's law of time warps dictates that time flow more slowly at the lower mouth than at the upper mouth. More slowly, that is, when compared along the path of gravity's intense pull: the dashed purple path through the external universe. I presume, for concreteness, that this has produced a one-hour lag so when compared through the external universe, the bottom clock shown in the figure is one hour behind the top clock. And this time lag is continuing to grow.

Fig. 30.3. Wormhole as a time machine.

Since there is only a tiny gravitational pull inside the wormhole, Einstein's law of time warps dictates that, as seen through the wormhole, time flows at essentially the same rate in the upper mouth as in the lower mouth. So there is no time lag when the clocks are compared through the wormhole. They are synchronized.

Suppose, further, for concreteness, that the distance from mouth to mouth in the external universe is short enough that you can traverse it in five minutes as measured by the clocks, and you can travel through the wormhole in one minute. Then this wormhole has already become a time machine. You leave the upper mouth at time 2:00 as measured by the clock there, and travel through the external universe to the lower mouth, arriving at 2:05 upper clock time and 1:05 lower clock time. You then make a one-minute trip upward through the wormhole, from lower mouth to upper. Since the clocks. are synchronized through the wormhole, you reach the upper mouth at time 1:06 as seen by both clocks. You arrive back at your starting point fifty-four minutes before your 2:00 departure, and you meet your younger self.

Some days earlier, when the time difference was much less, the wormhole was not yet a time machine. It became a time machine at the first moment when something, moving at the highest possible speed, the speed of light, was able to travel along your route and arrive back at the top mouth at the very moment it started out.

If that something is a particle of light (a photon), for example, then we began with one photon and we now have two, at the starting place and time. After those two make the trip, we have four at that same place and time, then eight, then sixteen, . . . ! There is a growing crescendo of energy coursing through the wormhole, perhaps enough that the energy's gravity destroys the wormhole at the very moment it is becoming a time machine.

It would seem easy to prevent this. Just shield the wormhole from photons. However, there is something you cannot shield out: quantum fluctuations of light with ultrahigh frequency—fluctuations that inevitably exist, according to the quantum laws (Chapter 26). In 1990, Sung-Won Kim (a postdoctoral student in my research group) and I used the quantum laws to compute the fate of such fluctuations. We found a growing explosion (Figure 30.4). We thought, at first, that the explosion was too weak to destroy the wormhole. The wormhole would become a time machine despite the explosion, we thought. Stephen Hawking convinced us otherwise. The fate of the explosion is controlled by the laws of quantum gravity, he convinced us. Only when those laws are well understood will we know for sure whether backward time travel is possible.

Stephen, however, was so convinced that the ultimate answer will

Fig. 30.4. Quantum fluctuations of light, traveling along the red path, build up into a crescendo explosion at the moment the wormhole is becoming a time machine.

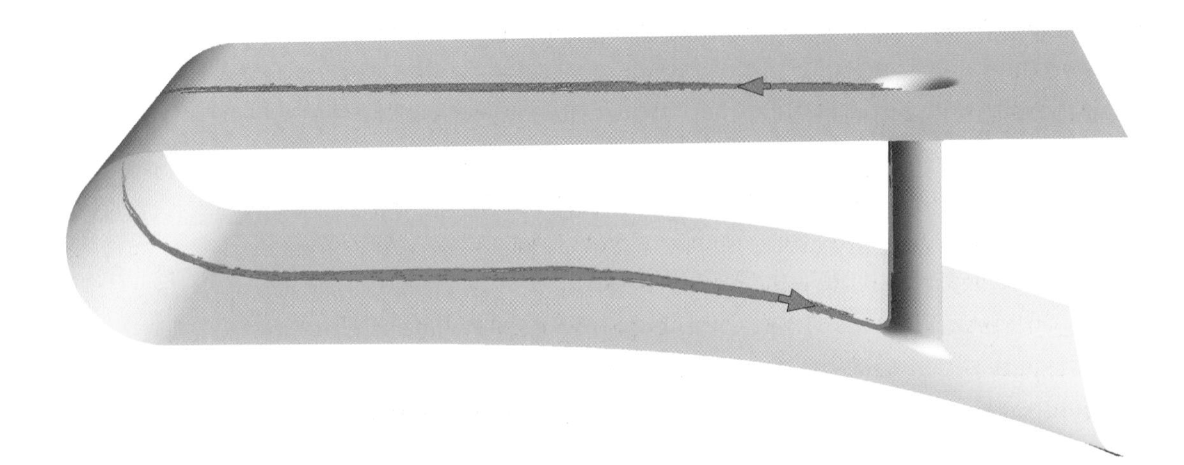

be no time machines, that he codified this in what he calls his "chronology protection conjecture": The laws of physics will always prevent backward time travel, thereby "keeping the universe safe for historians."

Many researchers have struggled, over the past twenty years, to prove or disprove Hawking's chronology protection conjecture. The bottom line today, I think, remains the same as in the early 1990s, when he and I were debating the issue: Only the laws of quantum gravity know for sure.

Time Travel with a Bulk

All this research and conclusions—educated guesses—are based on the laws of physics that prevail *if there is no bulk with a large fifth dimension*. What happens to time travel if a large bulk *does* exist, as in *Interstellar*?

We physicists find Einstein's relativistic laws so compelling that we suspect they hold in the bulk as well as in our brane. So Lisa Randall, Raman Sundrum, and others have extended his laws into the five-dimensional bulk by one simple step: adding a new dimension to space. That extension proceeds mathematically in a straightforward and beautiful manner, which makes us physicists think we may be on the right track. In my interpretation of the movie, Professor Brand uses this extension as a foundation for his equation and for his struggle to understand gravitational anomalies (Chapter 25).

If this speculative extension is correct, then time behaves fundamentally the same in the bulk as in our brane. In particular, objects and signals in the bulk, like those in our brane, can only move in one direction through locally measured time (local bulk time): toward the future. They cannot move backward, locally. If backward time travel is possible in the bulk, it can be achieved only by journeying out through the bulk's space and returning before the journey started while always moving forward in local bulk time. This is a bulk analog of the round trip in Figure 30.3.

Messaging Murph: My Physicist's Interpretation

This description of time underlies my physicist's interpretation of Cooper's messaging Murph.

Recall that the tesseract is an object whose faces have three space dimensions and interior has four. The interior is part of the bulk. Everything we see in the movie's tesseract scenes lies in the faces: Cooper, Murph, Murph's bedroom, the bedroom's extrusions, the world tubes of the book and watch—all lie in tesseract faces. We never see the tesseract's bulk interior. We can't see it, since light can't travel through four space dimensions, only three. However, gravity can do so.

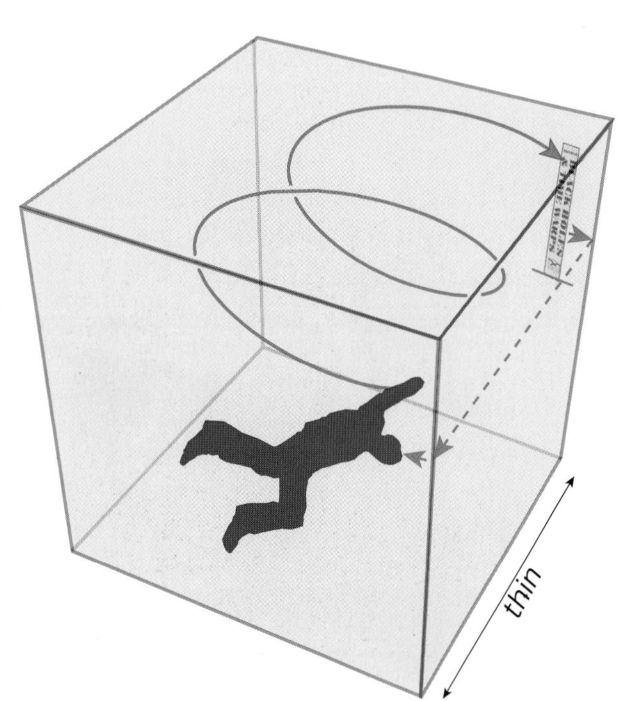

Fig. 30.5. A Cooper icon sees a book via the red dashed light ray and exerts a force on the book via a gravitational signal that spirals along the violet curve. I've suppressed one of our brane's spatial dimensions.

In my interpretation, when Cooper sees a book in Murph's bedroom, he does so via a light ray that travels in faces of the tesseract (for example, the red dashed ray in Figure 30.5). And when he pushes on a book's world tube, or on the world tube of the watch's second hand, he generates a gravitational signal (a gravitational wave in the bulk) that spirals into and through the tesseract's bulk interior, along the violet curve in Figure 30.5. The signal travels forward in local, bulk time, but backward in bedroom time, arriving before it started out.[2] It is this gravitational signal that pushes the book out of the bookcase and twitches the watch's second hand.

2 I can easily write down a mathematical description of spacetime warping that achieves this—a warping that bulk engineers could try to build to facilitate gravitational signals going forward in local bulk time, but backward relative to bedroom time; see the technical notes for this chapter, at the end of the book, especially Figure TN.1. Whether the bulk engineers could actually build this warping in practice depends on the laws of quantum gravity—laws that I don't know, but TARS discovers in Gargantua's singularity.

This is rather like one of my favorite Escher drawings, *Waterfall* (Figure 30.6). Downward in the drawing is analogous to the forward flow of bedroom time, and the flowing water is analogous to the forward flow of local time. A leaf on the water is carried forward with the water just like signals in the bulk are carried forward in local time.

When carried by water down the waterfall, the leaf is like the light ray from the book to Cooper: It travels not only forward in local time but also downward (forward in bedroom time). When carried along the aqueduct, the leaf is like the gravitational signal from Cooper to the book: it travels forward in local time but upward[3] (so backward in bedroom time).

How, in this interpretation, do I explain Amelia Brand's description of time as seen by beings in the bulk? "To Them time may be just another physical dimension. To Them the past might be a canyon They can climb into and the future a mountain They can climb up."

Fig. 30.6. *Waterfall.* [Drawing by M. C. Escher.]

Einstein's laws, extended into the bulk, tell us that local bulk time can't behave this way. Nothing in the bulk can go backward in local bulk time. However, when looking into our brane from the bulk, Cooper and bulk beings can and do see our brane's time (bedroom time) behave like Brand says. *As seen from the bulk,* "our brane's time can look like just another physical dimension," to paraphrase Brand. "Our brane's past looks like a canyon that Cooper can climb into [by traveling down the tesseract's diagonal channel], and our brane's future looks like a mountain that Cooper can climb up [by traveling up the tesseract's diagonal channel; Figure 29.14]."

This is my physicist's interpretation of Brand's words. And Chris interprets them similarly.

3 Via an optical illusion.

Touching Brand Across the Fifth Dimension

In *Interstellar,* with the quantum data safely in Murph's hands, Cooper's mission is finished. The tesseract, carrying him through the bulk, begins to close.

As it is closing, he sees the wormhole. And within the wormhole, he sees the *Endurance* on its maiden voyage to Gargantua. As he sweeps past the *Endurance,* he reaches out and gravitationally touches Brand across the fifth dimension. She thinks she has been touched by a bulk being. She has . . . by a being riding through the bulk in a rapidly closing tesseract. By an exhausted, older Cooper.

31

Lifting Colonies off Earth

Early in *Interstellar,* when Cooper first visits the NASA facility, he is shown a giant, cylindrical enclosure being constructed to carry thousands of humans into space and house them for many generations: a space colony. And he's told there are others being constructed elsewhere.

"How does it get off Earth?" Cooper asks the Professor. "Those first gravitational anomalies changed everything," the Professor replies. "Suddenly we knew that harnessing gravity was real. So I started working on the theory—and we started building this station."

At the end of *Interstellar* we see everyday life back on even keel, inside the colony, floating in space (Figure 31.1).

How did it get lifted into space? The key, of course, was the quantum data (in my scientist's interpretation, the *quantum gravity laws*) that TARS extracted from Gargantua's singularity (Chapters 26 and 28) and Cooper transmitted to Murph (Chapter 30).

In my interpretation, by discarding quantum fluctuations from those laws (Chapter 26), Murph learned the nonquantum laws that govern gravitational anomalies. And from those laws, she figured out how to control the anomalies.

As a physicist, I'm eager to know the details. Was Professor Brand

Fig. 31.1. Kids playing baseball inside the space colony, as seen by Cooper looking through a window. *[From* Interstellar, *used courtesy of Warner Bros. Entertainment Inc.]*

on the right track in the equations that covered his blackboards? (Chapter 25 and this book's page at Interstellar.withgoogle.com.) Did he really have half the answer, as Murph asserted before getting the quantum data? Or was he way off? Is the secret to anomalies and controlling gravity something completely different?

Perhaps a sequel to *Interstellar* will tell us. Christopher Nolan is a master of sequels; just watch his Batman trilogy.

But one thing seems clear. Murph must have figured out how to reduce Newton's gravitational constant G inside the Earth. Recall (Chapter 25) that the Earth's gravitational pull is given by Newton's inverse square law: $g = Gm/r^2$, where r^2 is the squared distance from the Earth's center, m is the mass of the Earth, and G is Newton's gravitational constant. Cut Newton's G in half and you reduce the Earth's gravity by two. Cut G by a thousand and you reduce the Earth's gravity by a thousand.

In my interpretation, with Newton's G reduced inside the Earth to, say, a thousandth its normal value for, say, an hour, rocket engines could lift the enormous colonies into space.

As a byproduct, in my interpretation the Earth's core—no longer compressed by the enormous weight of the planet above—must have sprung outward, pushing the Earth's surface upward. Gigantic earthquakes and tsunamis must have followed, wreaking havoc on Earth as the colonies soared into space, a terrible price for the Earth to pay on top of its blight-driven catastrophe. When Newton's *G* was restored to normal strength, the Earth must have shrunk back to its normal size, wreaking more earthquake and tsunami havoc.

But humanity was saved. And Cooper and ninety-four-year-old Murph were reunited. Then Cooper set out in search of Amelia Brand in the far reaches of the universe.

Some Parting Thoughts

Ⓣ

Every time I watch *Interstellar* and browse back through this book, I'm amazed at the enormous variety of science they contain. And the richness and beauty of that science.

More than anything, I'm moved by *Interstellar*'s underlying, optimistic message: We live in a universe governed by physical laws. By laws that we humans are capable of discovering, deciphering, mastering, and using to control our own fate. Even without bulk beings to help us, we humans are capable of dealing with most any catastrophe the universe may throw at us, and even those catastrophes we throw at ourselves—from climate change to biological and nuclear catastrophes.

But doing so, controlling our own fate, requires that a large fraction of us understand and appreciate science: How it operates. What it teaches us about the universe, the Earth, and life. What it can achieve. What its limitations are, due to inadequate knowledge or technology. How those limitations may be overcome. How we transition from speculation to educated guess to truth. How extremely rare are revolutions in which our perceived truth changes, yet how very important.

I hope this book contributes to that understanding.

WHERE CAN YOU LEARN MORE?

Chapter 1. A Scientist in Hollywood: The Genesis of *Interstellar*

For readers interested in the culture of Hollywood and the shifting sands of moviemaking, I highly recommend two books by my partner, Lynda Obst: *Hello, He Lied: & Other Truths from the Hollywood Trenches* (Obst 1996) and *Sleepless in Hollywood: Tales from the New Abnormal in the Movie Business* (Obst 2013).

Chapter 2. Our Universe in Brief

For an overview of our entire universe with lots of great pictures, and with connections to what you can see in the night sky with your naked eye, binoculars, and telescopes, see *Universe: The Definitive Visual Guide* (Rees 2005). Many good books have been written about what happened in our universe's earliest moments, its big-bang origin, and how the big bang may have gotten started. I particularly like *The Inflationary Universe* (Guth 1997); *Big Bang: The Origin of the Universe* (Singh 2004); *Many Worlds in One: The Search for Other Universes* (Vilenkin 2006); *The Book of Universes: Exploring the Limits of the Cosmos* (Barrow 2011); and Chapters 3, 14, and 16 of *From Eternity to Here: The Quest for the Ultimate Theory of Time* (Carroll 2011). For current research on the big bang, see the blog by Sean Carroll, *Preposterous Universe* (Carroll 2014) at http://www.preposterousuniverse.com/blog/.

Chapter 3. The Laws That Control the Universe

Richard Feynman, one of the great physicists of the twentieth century, gave a series of lectures for the general public in 1964 that delved deeply into the nature of the laws that control our universe. He wrote up his lectures in one of my favorite books of all time, *The Character of Physical Law* (Feynman 1965). For a more detailed, more up-to-date, and much longer book on the same topic, see *The Fabric of the Cosmos: Space, Time, and the Texture of Reality* (Greene 2004). Easier going, perhaps more fun, and equally deep is *The Grand Design* (Hawking and Mlodinow 2010).

Chapter 4. Warped Time and Space, and Tidal Gravity

For historical details on Einstein's concepts of warped time and space, their connection to tidal gravity, and his relativistic laws built on these concepts, see Chapters 1 and 2 of *Black Holes & Time Warps: Einstein's Outrageous Legacy* (Thorne 1994); and for a plethora of experiments that show Einstein was right, see *Was Einstein Right? Putting General Relativity to the Test* (Will 1993). *"Subtle Is the Lord . . .": The Science and the Life of Albert Einstein* (Pais 1982) is a biography of Einstein that focuses in depth on all of Einstein's contributions to science; it's much tougher going and much more scholarly than Thorne or Will. There are other, more comprehensive biographies of Einstein—I especially like *Einstein: His Life and Universe* (Isaacson 2007)—but no other biography treats Einstein's science with anything approaching the accuracy and detail of Pais.

Gravity from the Ground Up: An Introductory Guide to Gravity and General Relativity (Schutz 2003) is an in-depth discussion of gravity and its roles in our universe (both Newtonian gravity and Einstein's warped spacetime), written for the general reader. For the same material at the level of an advanced undergraduate physics or engineering student, I like the textbooks by James Hartle, *Gravity: An Introduction to Einstein's General Relativity* (Hartle 2003), and by Bernard Schutz, *A First Course in General Relativity* (Schutz 2009).

Chapter 5. Black Holes

For greater detail on black holes and how we came to know the things we think we know about them, I suggest *Gravity's Fatal Attraction: Black Holes in the Universe* (Begelman and Rees 2009), *Black Holes & Time Warps* (Thorne 1994), and a lecture that I gave in 2012 at Stephen Hawking's seventieth birthday party: http://www.ctc.cam.ac.uk/hawking70/multimedia_kt.html. Andrea Ghez describes her team's wonderful discoveries about the black hole at the center of our Milky Way Galaxy in a Ted talk at http://www.ted.com/speakers/andrea_ghez and on her team's website, http://www.galacticcenter.astro.ucla.edu.

Chapter 6. Gargantua's Anatomy

For properties of black holes that are featured in this chapter, see Chapter 7 of *Black Holes & Time Warps* (Thorne 1994), especially pp. 272–295; and at a more technical level, with equations, *Gravity: An Introduction to Einstein's General Relativity* (Hartle 2003). Also see the appendix *Some Technical Notes* in this book. For the shell of fire and the orbits of photons temporarily trapped in it, see Edward Teo's technical paper (Teo 2003).

Chapter 7. Gravitational Slingshots

For a discussion of gravitational slingshots at a modestly more technical level than mine, I recommend the Wikipedia article http://en.wikipedia.org/wiki/Gravity_assist. But don't believe what it says about slingshots around black holes. Its statement (as of July 4, 2014) that "if a spacecraft gets close to the Schwarzschild radius [horizon] of a black hole, space becomes so curved that slingshot orbits require more energy to escape than the energy that could be added by the black hole's motion" is just plain wrong. Indeed, you should always read Wikipedia with some cautious skepticism. In my experience, in areas where I am an expert, roughly 10 percent of Wikipedia's statements are wrong or misleading.

More reliable than Wikipedia for gravitational slingshots, but less comprehensive, is http://www2.jpl.nasa.gov/basics/grav/primer.php. A gravitational-slingshot video game has been developed in connection with *Interstellar*; see Game.InterstellarMovie.com.

For a somewhat technical discussion of the intermediate-mass black holes that I invoke for gravitational slingshots, see Chapter 4 of *Black Hole Astrophysics: The Engine Paradigm* (Meier 2012).

You can generate and explore complicated orbits around fast-spinning black holes, such as that in Figure 7.6, using a tool written by David Saroff and available at http://demonstrations.wolfram.com/3DKerrBlackHoleOrbits.

Chapter 8. Imaging Gargantua

Simulations of the gravitational lensing of star fields by black holes, similar to those that underlie *Interstellar*, have been carried out previously by a number of physicists and can be found on the web. Especially impressive are those by Alain Riazuelo; see www2.iap.fr/users/riazuelo/interstellar. See also the section on Chapter 28, below.

Paul Franklin's team and I plan to write several somewhat technical articles about the simulations that they carried out using the equations I gave them: the simulations underlying *Interstellar*'s images of Gargantua and its disk and the wormhole, and additional simulations that have revealed surprising things. You can access these articles on the web at http://arxiv.org/find/gr-qc.

Chapter 9. Disks and Jets

For in-depth discussions of quasars, accretion disks, and jets, see *Gravity's Fatal Attraction* (Begelman and Rees 2009), Chapter 9 of *Black Holes & Time Warps* (Thorne 1994), and at a more technical and more detailed level, *Black Hole Astrophysics* (Meier 2012). For the tidal disruption of stars by black holes and the resulting accretion disks, see the website of James Guillochon (who, with colleagues, was responsible for the simulations that underlie Figures 9.5 and 9.6): http://astrocrash.net/projects/tidal-disruption-of-stars/. For astrophysically realistic film clips of accretion disks and their jets, I recommend some by Ralf Kaehler (Stanford University) at http://www.slac.stanford.edu/~kaehler/homepage/visual

izations/black-holes.html, based on simulations by Jonathan C. McKinney, Alexander Tchekhovskoy, and Roger D. Blandford (McKinney, Tchekhovskoy, and Blandford 2012). For some images of accretion disks with Doppler shifts included as well as gravitational lensing, see the website of the astrophysicist Avery Broderick, http://www.science.uwaterloo.ca/~abroderi/Press/. The simulations that underlie Gargantua's accretion disk in *Interstellar* (for example, Figure 9.9) will be described in one or more articles to appear at http://arxiv.org/find/gr-qc.

Chapter 10. Accident Is the First Building Block of Evolution

I don't know any nontechnical discussions of the simulations that show the star density near a massive black hole growing, rather than decreasing. For a technical discussion and analysis, see Chapter 7 of *Dynamics and Evolution of Galactic Nuclei* (Merritt 2013), particularly Figure 7.4.

Chapter 11. Blight

If you watch the daily science news, or just observe the world around you, you'll see examples of the kinds of scenarios that my biologist colleagues describe in this chapter—mild examples, thus far, fortunately; not catastrophic examples. A recent one is the amazing jump of a lethal virus from plants to honeybees, http://blogs.scientificamerican.com/artful-amoeba/2014/01/31/suspicious-virus-makes-rare-cross-kingdom-leap-from-plants-to-honeybees; this was a far bigger jump than that from okra to corn in *Interstellar*, but a far less lethal pathogen. Another example is the rapid demise of tree species once dominant on the American scene: not only the American chestnut tree mentioned by Meyerowitz in Chapter 11, but the American elm tree, http://landscaping.about.com/cs/treesshrubs/a/american_elms.htm, and the giant pine trees around my cabin on Palomar Mountain, near the 200-inch telescope.

Chapter 12. Gasping for Oxygen

The cycling of oxygen between the breathable oxygen molecule O_2, and carbon dioxide CO_2, and also (more slowly) other forms, is called the Earth's "oxygen cycle." Google it. The cycling of carbon between

CO_2 in the atmosphere, plants (dead and alive), and also (much more slowly) other forms such as coal, oil, and kerogen, is called the "carbon cycle." Google it, too. Obviously these cycles are coupled; they influence each other. They are the foundation for Chapter 13.

Chapter 13. Interstellar Travel

Exoplanets (planets beyond our solar system) are being discovered at a furious pace. Nearly complete catalogs, updated daily, are at http://exoplanet.eu and http://exoplanets.org. A catalog of exoplanets that could be habitable is at http://phl.upr.edu/hec. For the human side and history of the search for exoplanets and life beyond the solar system, see *Mirror Earth: The Search for Our Planet's Twin* (Lemonick 2012) and *Five Billion Years of Solitude: The Search for Life Among the Stars* (Billings 2013); for technical and scientific details, see *The Exoplanet Handbook* (Perryman 2011). *Confessions of an Alien Hunter: A Scientist's Search for Extraterrestrial Intelligence* (Shostak 2009) is an excellent description of the search for extraterrestrial intelligence (SETI) via radio signals from beyond Earth and by other methods.

For information about technologies that we humans could pursue in our quest for interstellar travel, I suggest http://en.wikipedia.org/wiki/Interstellar_travel and http://fourthmillenniumfoundation.org. The astronaut Mae Jemmison is spearheading a quest to send humans beyond the solar system in the next century; see http://100yss.org. A lot of nonsense is written about interstellar travel via warp drives and wormholes. The technology of this century and likely the next few is incapable of any realistic effort in this direction, unless some far more advanced civilization provides us with the necessary spacetime warps, as in *Interstellar.* So don't waste your time reading articles and claims about us humans producing strong enough warps for interstellar travel in your lifetime or that of your great-grandchildren.

Chapter 14. Wormholes

For greater detail on wormholes, I especially recommend *Lorentzian Wormholes: From Einstein to Hawking* (Visser 1995), despite its being nearly twenty years old. I also recommend the last chapter of *Black Holes & Time Warps* (Thorne 1994), Chapter 9 of *Time Travel and Warp Drives* (Everett and Roman 2012), and Chapter 8 of *Black Holes, Wormholes, and Time Machines* (Al-Khalili 2012). For an up-to-date discussion of the exotic matter required to hold a wormhole open, see Chapter 11 of *Time Travel and Warp Drives* (Everett and Roman 2012).

Chapter 15. Visualizing *Interstellar*'s Wormhole

Paul Franklin's team and I give much greater detail about our work on wormhole visualization in one or more articles that we plan to make available on the web at http://arxiv.org/find/gr-qc.

Chapter 16. Discovering the Wormhole: Gravitational Waves

For up-to-date information about LIGO and the search for gravitational waves, see the website of the LIGO Scientific Collaboration, http://www.ligo.org, especially the "News" and "Magazine" sections; also the LIGO Laboratory's website http://www.ligo.caltech.edu, and also Kai Staats's 2014 movie at http://www.space.com/25489-ligo-a-passion-for-understanding-complete-film.html. On the web you can also find a number of pedagogical lectures by me about gravitational waves and the warped side of the universe, for example my three "Pauli Lectures" at http://www.multimedia.ethz.ch/speakers/pauli/2011, which should be watched in the opposite order to their listing (that is, from the bottom, upward); and at a moderately technical level, http://www.youtube.com/watch?v=Lzrlr3b5aO8. For movies of black-hole collisions and the gravitational waves they emit, based on the SXS team's simulations, see http://www.blackholes.org/explore2.html.

There are no up-to-date books about gravitational waves for the general reader, but I do recommend *Einstein's Unfinished Symphony: Listening to the Sounds of Space-Time* (Bartusiak 2000), which is not extremely out of date. For the history of research on gravitational waves from Einstein onward, see *Traveling at the Speed of Thought: Einstein and the Quest for Gravitational Waves* (Kennefick 2007).

Chapter 17. Miller's Planet

In this chapter I make a large number of claims about Miller's planet: its orbit, its rotation (it always keeps the same face toward Gargantua except for rocking), Gargantua's tidal forces that deform it and make it rock; and Gargantua's whirl of space that it experiences and how the whirl influences inertia, centrifugal forces, and the speed-of-light speed limit. These claims are all supported by Einstein's relativistic laws of physics, his general relativity. I don't know of any books or articles or lectures for nonspecialists that discuss and explain these things, for a planet orbiting close to a spinning black hole, except my Chapter 17. Readers at the advanced undergraduate level may try to check my claims using concepts and equations in Hartle's textbook, *Gravity: An Introduction to Einstein's General Relativity* (Hartle 2003).

The questions I raise in the section "Past History of Miller's Planet" do not require much relativistic physics. They can be answered almost entirely with Newton's laws of physics, and the best places to seek relevant information are books and websites that deal with geophysics or the physics of planets and their moons.

Chapter 18. Gargantua's Vibrations

For a description of Bill Press's discovery that black holes can vibrate and Saul Teukolsky's deduction of the equations that govern those vibrations, see pp. 295–299 of *Black Holes & Time Warps* (Thorne 1994). The technical article about black-hole vibrations and their ring-down that underlies both Figure 18.1 and Romilly's data set is Yang et al. (2013) by Huan Yang, Aaron Zimmerman, and their colleagues.

Chapter 21. The Fourth and Fifth Dimensions

For more detail on the unification of space and time, see pp. 73–79 of *Black Holes & Time Warps* (Thorne 1994). For the superstring breakthrough by John Schwarz and Michael Green and how that forced physicists to embrace a bulk with extra dimensions, see *The Elegant Universe: Superstrings, Hidden Dimensions, and the Quest for the Ultimate Theory* (Greene 2003).

Chapter 22. Bulk Beings

For a highly rated, animated movie of Edwin A. Abbott's *Flatland* (Abbott 1884), see *Flatland: The Film* (Ehlinger 2007). For extensive discussions of the mathematics underlying *Flatland* and the story's connections to nineteenth-century English society, see *The Annotated Flatland: A Romance of Many Dimensions* (Stewart 2002). For visual insights into the fourth space dimension, see *The Visual Guide to Extra Dimensions, Volume 1: Visualizing the Fourth Dimension, Higher-Dimensional Polytopes, and Curved Hypersurfaces* (McMullen 2008).

Chapter 23. Confining Gravity

For much of the content of this chapter, I recommend *Warped Passages: Unraveling the Mysteries of the Universe's Hidden Dimensions* (Randall 2006). This is a thorough discussion of modern physicists' ideas and predictions about the bulk and its extra dimensions, written by Lisa Randall who, with Raman Sundrum, discovered that AdS warping can confine gravity near our brane (Figures 23.4 and 23.6). The idea of an AdS layer and sandwich, which I rediscovered, was first proposed and discussed in a technical paper by Ruth Gregory, Valery A. Rubakov, and Sergei M. Sibiryakov (Gregory, Rubakov, and Sibiryakov 2000), and the AdS sandwich was shown to be unstable in a technical paper by Edward Witten (Witten 2000).

Chapter 24. Gravitational Anomalies

For the history of the anomalous precession of Mercury's orbit and the search for the planet Vulcan, I recommend a scholarly treatise by science historian N. T. Roseveare, *Mercury's Perihelion from Le Verriere to Einstein* (Roseveare 1982), and also the more readable but less comprehensive account by astronomers Richard Baum and William Sheehan, *In Search of the Planet Vulcan: The Ghost in Newton's Clockwork Universe* (Baum and Sheehan 1997).

For the discovery of evidence for dark matter in our universe and the current search for dark matter, I recommend a highly readable book, *The Cosmic Cocktail: Three Parts Dark Matter* (Freeze 2014), by one of the leading researchers in this quest, Katherine Freeze.

For the anomalous acceleration of the universe's expansion and the dark energy that presumably causes it, I recommend the last chapter of *The Cosmic Cocktail* (Freeze 2014) and also *The 4% Universe: Dark Matter, Dark Energy, and the Race to Discover the Rest of Reality* (Panek 2011).

Chapter 25. The Professor's Equation

The ideas that Newton's gravitational constant *G* might change from place to place and time to time, and might be controlled by some sort of nongravitational field, were hot topics in the Princeton University physics department when I was a PhD student there in the early 1960s. These ideas had been proposed by Princeton's Professor Robert H. Dicke and his graduate student Carl Brans in connection with their "Brans-Dicke theory of gravity" (Chapter 8 of *Was Einstein Right?* [Will 1993]), an interesting alternative to Einstein's general relativity. For a brief personal memoir about this, see "Varying Newton's Constant: A Personal History of Scalar-Tensor Theories" in *Einstein Online* (Brans 2010). The Brans-Dicke theory has motivated a number of experiments that searched for varying *G*, but no convincing variations were ever found; see, for example, Chapter 9 of *Was Einstein Right?* (Will 1993). These ideas and experiments motivated my interpretation of some of *Interstellar*'s gravitational anomalies and how to control them: bulk fields control the strength of *G* and make it vary.

The Professor's equation, shown on his blackboard in Figure 25.6, builds on these ideas. It also incorporates Einstein's relativistic laws (general relativity), extended into the bulk's fifth dimension, which are laid out in a technical review article by Roy Maartens and Koyama Kazuya (Maartens and Kazuya 2010), and it incorporates a branch of mathematics called the "calculus of variations"; see, for example, http://en.wikipedia.org/wiki/Calculus_of_variations. For a few technical details about the Professor's equation, see the appendix *Some Technical Notes*.

Chapter 26. Singularities and Quantum Gravity

For a first foray into quantum fluctuations and quantum physics more generally, I recommend *The Ghost in the Atom: A Discussion of the Mysteries of Quantum Physics* (Davies and Brown 1986). I don't know any articles or books for nonphysicists about the quantum behavior of human-sized objects such as LIGO's mirrors; at a technical level, I discuss this in the second half of my third Pauli lecture (the one listed first) at http://www.multimedia.ethz.ch/speakers/pauli/2011. In John Wheeler's autobiography, he discusses how he came up with the idea of quantum foam (Chapter 11 of *Geons, Black Holes and Quantum Foam: A Life in Physics* [Wheeler and Ford 1998]).

In Chapter 11 of *Black Holes & Time Warps* (Thorne 1994) I discuss what was known in 1994 about the interiors of black holes, and how we came to know it—including the BKL singularity and its dynamics; quantum gravity's control of the singularity's core and its connection to quantum foam; and the infalling singularity (mass-inflation singularity), which had only recently been discovered by Erik Poisson and Werner Israel (Poisson and Israel 1990) and was not yet fully understood. The upflying singularity was discovered so recently that there is not yet any detailed discussion of it for nonphysicists; the technical discovery article is Marolf and Ori (2013) by Donald Marolf and Amos Ori. Matthew Choptuik's discovery that tiny, transient naked singularities are possible was announced and explained in his technical article (Choptuik 1993).

Chapter 27. The Volcano's Rim

The volcano-like surface that underlies much of this chapter (Figures 27.3, 27.5, and 27.9) can be described with elementary physics equations, as can the *Endurance*'s trajectory, the trajectory's instability on the rim, and the *Endurance*'s launch toward Miller's planet. See the appendix *Some Technical Notes*.

Chapter 28. Into Gargantua

In the Prologue of *Black Holes & Time Warps* (Thorne 1994), I describe, in much greater detail than here, what it would look like and feel like to fall through a black hole's horizon, both as seen and felt by the infalling person and as seen by someone else outside the black hole. And I describe how the look and feel are influenced by the mass of the black hole and by its spin.

Andrew Hamilton has constructed a "Black Hole Flight Simulator" for computing what it looks like to fall into a nonspinning black hole. His computations are similar to those done for *Interstellar* by Paul Franklin's team (Chapters 8, 9, and 15), but preceded *Interstellar* by many years. Andrew has used his simulator to produce a remarkable set of film clips that can be found on his website, http://jila.colorado.edu/~ajsh/insidebh, and in planetariums around the world (see http://www.spitzinc.com/fulldome_shows/show_blackholes).

Andrew's film clips differ from what we see in *Interstellar* in several ways: First, for pedagogical purposes Andrew sometimes paints a grid of lines on the black hole's horizon (there is no such grid for real black holes and none in *Interstellar*), and when he does so, he also replaces the star that imploded to form the black hole by a "past horizon."[1] Second, in his "Journey into a Realistic Black Hole," http://jila.colorado.edu/~ajsh/insidebh/realistic.html, Andrew endows the hole with a jet and an accretion disk. Gas from the disk falls into and through the horizon, and that infalling gas dominates what the camera sees at and beneath the horizon. In *Interstellar,* by contrast, there is no jet, and the accretion disk is so anemic that it is not currently sending any of its gas into and through the horizon, so the hole's interior looks rather dark. However, in *Interstellar* Cooper encounters a dim fog of light and white flakes from stuff that fell in before him. These are not the result of simulations, but instead were put in by hand by the Double Negative artists.

1 Stated more precisely and more technically, he has his camera fall into the maximally extended Schwarzschild solution or Reissner-Nordstrom solution of Einstein's equations instead of into a black hole.

Chapter 29. The Tesseract

When Christopher Nolan told me he was going to use a tesseract in *Interstellar,* I was delighted. At age thirteen I read about tesseracts in Chapter 4 of George Gamow's marvelous book *One, Two, Three, . . . Infinity* (Gamow 1947), and that had a major role in making me want to become a theoretical physicist. You can find a detailed discussion of tesseracts in *The Visual Guide to Extra Dimensions* (McMullen 2008). Christopher Nolan's complexified tesseract is unique; there is not yet any public discussion of it anywhere, except in this book and others connected to the movie *Interstellar.*

In Madeleine L'Engle's classic science fantasy novel for children, *A Wrinkle in Time* (L'Engle 1962), children travel via a tesseract—they "tesser"—to find their father. My own interpretation of this is a journey through the bulk, riding in the face of a tesseract, like my interpretation of Cooper's trip from Gargantua's core to Murph's bedroom, Figure 29.4.

Chapter 30. Messaging the Past

For physicists' current understanding of backward time travel in four spacetime dimensions without a bulk, see the last chapter of *Black Holes & Time Warps* (Thorne 1994), the chapters by Hawking, Novikov, and me in *The Future of Spacetime* (Hawking et al. 2002), and *Time Travel and Warp Drives* (Everett and Roman 2012). These are all by physicists who have contributed in major ways to the theory of time travel. For a historical account of modern research on time travel, see *The New Time Travelers: A Journey to the Frontiers of Physics* (Toomey 2007). For a comprehensive discussion of time travel in physics, in metaphysics, and in science fiction, see *Time Machines: Time Travel in Physics, Metaphysics and Science Fiction* (Nahin 1999). *From Eternity to Here: The Quest for The Ultimate Theory of Time* (Carroll 2011) is a wonderful discussion of almost everything physicists know, or speculate, about the nature of time.

I don't know any good books or articles, for general readers, about time travel when our universe is a brane that lives in a higher dimensional bulk; but as I discuss in Chapter 30, Einstein's laws extended to higher dimensions give basically the same predictions as without a bulk.

For some technical details of Cooper's sending messages backward in time to Murph, see the appendix *Some Technical Notes*.

Chapter 31. Lifting Colonies off Earth

For Murph's method (reducing G) for lifting the colonies off Earth, in my interpretation of *Interstellar*, see my remarks about Chapter 25, above.

In the early 1960s, when I was a PhD student at Princeton University, one of my physics professors, Gerard K. O'Neill, was embarking on an ambitious feasibility study for colonies in space, colonies somewhat like the one we see at the end of *Interstellar*. His study, augmented by a NASA study that he led, resulted in a remarkable book, *The High Frontier: Human Colonies in Space* (O'Neill 1978), which I highly recommend. But do pay attention to the book's introduction by Freeman Dyson, which discusses why O'Neill's dream of space colonies in his lifetime was shattered, but envisions them in the more distant future.

SOME TECHNICAL NOTES

The laws of physics that govern our universe are expressed in the language of mathematics. For readers comfortable with math, I write down a few formulas that come from the physical laws and show how I used them to deduce some things in this book. Two numbers that appear frequently in my formulas are the speed of light, $c = 3.00 \times 10^8$ meters/second, and Newton's gravitational constant, $G = 6.67 \times 10^{-11}$ meters^3/kilogram/second^2. I use scientific notation so 10^8 means 1 with eight zeros after it, 100,000,000 or a hundred million, and 10^{-11} means 0.[ten zeros]1, that is, 0.00000000001. I don't aspire to accuracy any higher than 1 percent, so I show only two or three digits in my numbers, and when a number is very poorly known, only one digit.

Chapter 4. Warped Time and Space, and Tidal Gravity

The simplest, quantitative form of Einstein's law of time warps is this: Place two identical clocks near each other, and at rest with respect to each other, separated from each other along the direction of the gravitational pull that they feel. Denote by R the fractional difference in their ticking rates, by D the distance between them, and by g the acceleration of gravity that they feel (which points from the one that ages the fastest to the one that ages the slowest). Then Einstein's law says that $g = Rc^2/D$. For the Pound-Rebca experiment in the Harvard tower, R was 210 picoseconds in one day, which is 2.43×10^{-15}, and the tower height D was 73 feet (22.3 meters). Inserting these into Einstein's law, we deduce $g = 9.8$ meters/second^2, which indeed is the gravitational acceleration on Earth.

Chapter 6. Gargantua's Anatomy

For a black hole such as Gargantua that spins extremely fast, the horizon's circumference C in the hole's equatorial plane is given by the formula $C = 2\pi GM/c^2 = 9.3\ (M/M_{sun})$ kilometers. Here M is the hole's mass, and $M_{sun} = 1.99 \times 10^{30}$ kilograms is the Sun's mass. For a very slowly spinning hole, the circumference is twice this size. The horizon's radius is defined to be this circumference divided by 2π: $R = GM/c^2 = 1.48 \times 10^8$ kilometers for Gargantua, which is very nearly the same as the radius of the Earth's orbit around the Sun.

The reasoning by which I deduce Gargantua's mass is this: The mass m of Miller's planet exerts an inward gravitational acceleration g on the planet's surface given by Newton's inverse square law: $g = Gm/r^2$, where r is the planet's radius. On the faces of the planet farthest from Gargantua and nearest it, Gargantua's tidal gravity exerts a stretching acceleration (difference of Gargantua's gravity between the planet's surface and its center a distance r away) given by $g_{tidal} = (2GM/R^3)r$. Here R is the radius of the planet's orbit around Gargantua, which is very nearly the same as the radius of Gargantua's horizon. The planet will be torn apart if this stretching acceleration on its surface exceeds the planet's own inward gravitational acceleration, so g_{tidal} must be less than g: $g_{tidal} < g$. Inserting the formulas above for g, g_{tidal}, and R, and expressing the planet's mass in terms of its density ρ as $m = (4\pi/3)r^3\rho$, and performing some algebra, we obtain $M < \sqrt{3}c^3/\sqrt{2\pi G^3\rho}$. I estimate the density of Miller's planet to be $\rho = 10{,}000$ kilograms/meter^3 (about that of compressed rock), from which I obtain $M < 3.4 \times 10^{38}$ kilograms for Gargantua's mass, which is about the same as 200 million suns—which in turn I approximate as 100 million suns.

Using Einstein's relativistic equations, I have deduced a formula that connects the slowing of time on Miller's planet, S = one hour/(seven years) $= 1.63 \times 10^{-5}$ to the fraction α by which Gargantua's spin rate is less than its maximum possible spin: $\alpha = 16S^3/(3\sqrt{3})$. This formula is correct only for very fast spins. Inserting the value of S, we obtain $\alpha = 1.3 \times 10^{-14}$; that is, Gargantua's actual spin is less than its maximum possible spin by about one part in a hundred trillion.

Chapter 8. Imaging Gargantua

The equations that I gave to Oliver James at Double Negative, for the orbital motion of light rays around Gargantua, are a variant of those in Appendix A of Levin and Perez-Giz (2008). Our equations for the evolution of bundles of rays are a variant of those in Pineult and Roeder (1977a) and Pineult and Roder (1977b). In several papers that we'll make available at http://arxiv.org/find/gr-qc, Paul Franklin's team and I give the specific forms of our equations and discuss details of their implementation and the simulations that resulted.

Chapter 12. Gasping for Oxygen

Here are the calculations that underlie my statements in Chapter 13. They are a nice example of how a scientist makes estimates. These numbers are very approximate; I quote them accurate to only one digit.

The mass of the Earth's atmosphere is 5×10^{18} kilograms, of which about 80 percent is nitrogen and 20 percent is molecular oxygen, O_2—that is, 1×10^{18} kilograms of O_2. The amount of carbon in undecayed plant life (called "organic carbon" by geophysicists) is about 3×10^{15} kilograms, with roughly half in the oceans' surface layers and half on land (Table 1 of Hedges and Keil [1995]). Both forms get oxidized (converted to CO_2) in about thirty years on average. Since CO_2 has two oxygen atoms (that come from the atmosphere) and just one carbon atom, and the mass of each oxygen atom is 16/12 that of a carbon atom, the oxidization of all this carbon, after all plants die, would eat up $2 \times 16/12 \times (3 \times 10^{15}$ kilograms$) = 1 \times 10^{16}$ kilograms of O_2, which is 1 percent of the atmosphere's oxygen.

For evidence of sudden overturns of the Earth's oceans and the theory of how they might be produced, see Adkins, Ingersoll, and Pasquero (2005). The standard estimate of the amount of organic carbon in sediments on the ocean bottoms that might be brought to the surface by such an overturn focuses on an upper sedimentary layer that is mixed by ocean currents and animal activity. This mixed layer's carbon content is the product of an estimated rate of deposit of carbon into the sediments (about 10^{11} kilograms per year) and the average time it takes for its carbon to be oxidized by oxygen from ocean water (1000 years), giving 1.5×10^{14} kilograms, one-twentieth of that on land and in ocean surface layers (Emerson and Hedges 1988, Hedges and Keil 1995). However: (i) The estimated deposition rate could be wrong by a huge amount; for example, Baumgart et al. (2009), relying on extensive measurements, estimate a deposition rate in the Indian Ocean off Java and Sumatra that is uncertain by a factor of fifty and, extrapolated to the whole ocean could give as much as 3×10^{15} kilograms of carbon in the mixed layer (the same as on land and in the ocean's surface layers). (ii) A substantial fraction of the deposited carbon could sink into a lower layer of sediment that does not get mixed into contact with seawater and oxidized except possibly during sudden ocean overturns. The last overturn is thought to have been during the most recent ice age, about 20,000 years ago—twenty times longer than the oxidation time in the mixed layer. So the unmixed layer could have twenty times more organic carbon than the mixed layer, and as much as twenty times that on land and in the ocean's surface. If brought to the ocean surface by a new overturn and there oxidized, this is nearly enough to leave everyone gasping for oxygen and dying of CO_2 poisoning; see the end of Chapter 12. Thus such a scenario is conceivable, though highly unlikely.

Chapter 15. Visualizing *Interstellar*'s Wormhole

Christopher Nolan chose several kilometers for the diameter of *Interstellar*'s wormhole. The wormhole's angular diameter as seen from Earth, in radians, is this diameter divided by its distance from Earth, which is about 9 astronomical units or 1.4×10^9 kilometers (the radius of Saturn's orbit). Therefore, the wormhole's angular diameter is about (2 kilometers)/(1.4×10^9 kilometers) = 1.4×10^{-9} radians, which is 0.0003 arc-seconds. Radio telescopes routinely achieve this angular resolution using transworld interferometry. Optical telescopes on the ground using a technique called "adaptive optics," and the Hubble space telescope in space, achieve angular resolutions a hundred times worse than this in 2014. Interferometry between twin Keck telescopes in Hawaii in 2014 can achieve a resolution ten times worse than the wormhole's angular diameter, and it is very plausible that in the era of *Interstellar* optical interferometry between more widely spaced optical telescopes will make possible resolutions better than the wormhole's 0.0003 arc-seconds.

Chapter 17. Miller's Planet

If you are familiar with Newton's gravitational laws in mathematical form, then you may find it interesting to explore a modification of them by the astrophysicists Bohdan Paczynski and Paul Wiita (Paczynski and Wiita 1980). In this modification, the gravitational acceleration of a nonspinning black hole is changed from Newton's inverse square law, $g = GM/r^2$ to $g = GM/(r - r_h)^2$. Here M is the hole's mass, r is the radius outside the hole at which the acceleration g is felt, and $r_h = 2GM/c^2$ is the radius of the nonspinning hole's horizon. This modification is a surprisingly good approximation to the gravitational acceleration predicted by general relativity.[1] Using this modified gravity, can you give a quantitative version of Figure 17.2[2] and deduce the radius of the orbit of Miller's planet? Your result will be only roughly correct, because the Paczynski-Wiita description of Gargantua's gravity fails to take account of the dragging of space into a whirling motion by the black hole's spin.

Chapter 25. The Professor's Equation

The meaning of the various mathematical symbols that appear in the Professor's equation (Figure 25.6) is explained on his other fifteen blackboards, which can be found on the web at this book's page at Interstellar.withgoogle.com. His equation expresses an "Action" S (the classical limit of a "quantum effective action") as an integral over "Lagrangian" functions $\mathcal{L}$. These Lagrangians involve the spacetime geometries ("metrics") of the five-dimensional bulk and our four-dimensional

1 This Paczynski-Wiita modification of gravity was used in developing the black hole's influence on spacecraft orbits for a gravitational-slingshot video game associated with *Interstellar*; see Game.InterstellarMovie.com.

2 For a related calculation, see the technical notes for Chapter 27, below.

brane, and also involve a set of fields that live in the bulk (denoted Q, σ, λ, ξ, and φ^i), and also "standard model fields" that live in our brane (including the electric and magnetic fields). The fields and spacetime metrics are to be varied, seeking an extremum (maximum or minimum or saddle point) of the Action **S**. The conditions that produce an extremum are a set of "Euler-Lagrange" equations that control the evolutions of the fields. This is a standard procedure in the calculus of variations. The Professor and Murph make guesses for a list of unknown bulk fields φ^i and unknown functions $U(Q)$, $H_{ij}(Q^2)$, $\mathcal{M}$ (standard model fields), and unknown constants W_{ij} that appear in the Lagrangian. In Figure 25.9 you see me writing a list of their guesses on the blackboard. Then for each set of guesses, they vary the fields and spacetime geometries, deduce the Euler-Lagrange equations, and then explore in computer simulations those equations' predictions for the gravitational anomalies.

Chapter 27. The Volcano's Rim

This note is for readers who are familiar with the mathematical description of Newton's laws of gravity and the conservation of energy and angular momentum. I challenge you to deduce the following formula for the volcano-like surface from (i) the Paczynski-Wiita approximate formula for Gargantua's gravitational acceleration, $g = GM/(r - r_h)^2$ (see the technical notes for Chapter 17, above) and (ii) the conservation laws for energy and angular momentum. The formula, using the notation of the technical notes for Chapter 17 plus L for the *Endurance*'s angular momentum (per unit mass), is

$$V(r) = -\frac{GM}{r - r_h} + \frac{1}{2}\frac{L^2}{r^2}.$$

The first term is the *Endurance*'s gravitational energy (per unit mass), the second is its circumferential kinetic energy, and the sum of $V(r)$ and the radial kinetic energy $v^2/2$ (with v its radial velocity) is equal to the *Endurance*'s conserved total energy (per unit mass). The rim of the volcano is at the radius r where $V(r)$ is a maximum. I challenge you, using these equations and ideas, to prove my claims, in Chapter 27, about the *Endurance*'s trajectory, the trajectory's instability on the rim of the volcano, and its launch toward Edmunds' planet.

Chapter 30. Messaging the Past

In the bulk as well as in our brane, the locations in spacetime, to which messages and other things can travel, are controlled by the law that nothing can travel faster than light. We physicists use spacetime diagrams to explore the consequences of this law. We draw spacetime diagrams in which, at each event, there is a "future light cone." Light travels outward from that event along the light cone; everything else, moving slower than light, travels from that event either along or inside the cone. See, for example, *Gravity: An Introduction to Einstein's General Relativity* (Hartle 2003).

Figure TN.1 shows the pattern of future light cones inside and on faces of the tesseract, in my interpretation of *Interstellar.* (It is the mathematical description of spacetime warping that I refer to in footnote 1 of Chapter 30. Physicists call this pattern of light cones "the causal structure of spacetime" inside the tesseract.) Figure TN.1 also shows the world line (violet curve) of the gravitational-wave message (force) sent by Cooper through the tesseract's interior to Murph's bedroom; and the world line (red dashed line) of the

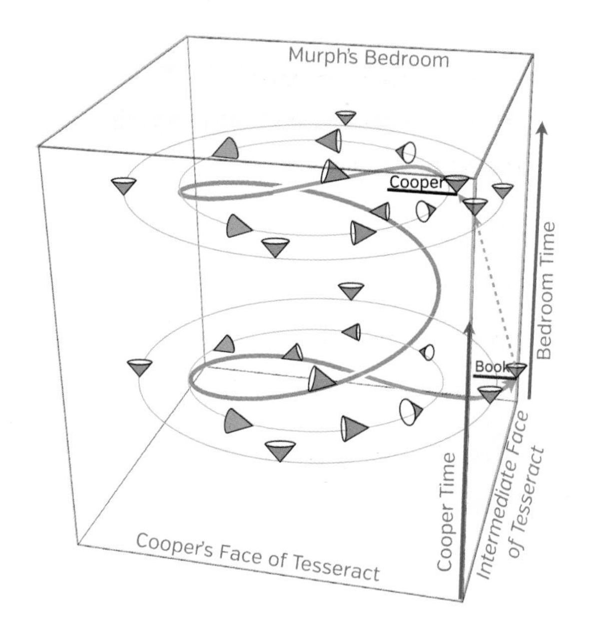

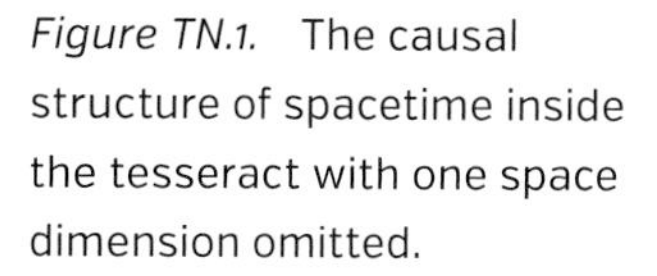
Figure TN.1. The causal structure of spacetime inside the tesseract with one space dimension omitted.

light ray from the bedroom through tesseract faces, by which Cooper sees the bedroom. This is a spacetime version of the purely spatial diagram in Figure 30.5.

Can you understand from this diagram how it is that the gravitational-wave message travels at the speed of light, yet moves backward relative to bedroom time and Cooper's time? And can you understand how, by contrast, the light ray travels at the speed of light and moves forward relative to bedroom time and Cooper's time? Compare with our discussion of Escher's drawing, Figure 30.6.

ACKNOWLEDGMENTS

For welcoming me into Hollywood and teaching me so much about that remarkable world, I thank, first and foremost, my partner Lynda Obst; and also Christopher Nolan, Emma Thomas, Jonathan Nolan, Paul Franklin, and Steven Spielberg.

I thank Lynda for the friendship and collaboration that gave birth to the treatment from which *Interstellar* sprang, and for guiding *Interstellar* through its trials and tribulations until it landed in the remarkable hands of Christopher Nolan, who transformed it so greatly.

For welcoming me into the visual-effects world and giving me the opportunity to lay foundations for visualizing *Interstellar*'s wormhole and the black hole Gargantua and its accretion disk, I thank Paul Franklin, Oliver James, and Eugénie Von Tunzelmann; and for collaborating tightly with me on those foundations, I thank Oliver and Eugénie.

For wise comments and suggestions on the manuscript of this book, I'm grateful to Lynda Obst, Jeff Shreve, Emma Thomas, Christopher Nolan, Jordan Goldberg, Paul Franklin, Oliver James, Eugénie Von Tunzelmann, and Carol Rose. For their dogged commitment to accuracy and consistency in every line of the manuscript, I thank Leslie Huang and Don Rifkin. For crucial assistance and/or advice about figures, I thank Jordan Goldberg, Eric Lewy, Jeff Shreve, Julia Druskin, Joe Lops, Lia Halloran, and Andy Thompson. For crucial assistance in getting permission for use of figures, I thank Pat Holl.

And for making the book a reality, I'm grateful to Drake McFeely, Jeff Shreve, Amy Cherry, and my Hollywood attorneys Eric Sherman and Ken Ziffren (yes, most everyone who works in Hollywood has to have an attorney or agent; even a scientist on the fringes).

And for her patience and support throughout this adventure, I'm grateful to my wife and life partner, Carolee Winstein.

FIGURE CREDITS

Following figures © Warner Bros. Entertainment Inc.: 1.2, 3.3, 3.4, 3.6, 5.6, 8.1, 8.5, 8.6, 9.7, 9.9, 9.10, 9.11, 11.1, 14.9, 15.2, 15.4, 15.5, 17.5, 17.9, 18.1, 19.2, 19.3, 20.1, 20.2, 24.5, 25.1, 25.7, 25.8, 25.9, 27.8, 28.3, 29.8, 29.14, 30.1, 31.1

Following figures © Kip Thorne: 2.4, 2.5, 3.2, 3.5, 4.3, 4.4, 4.8, 4.9, 5.1, 5.2, 5.3, 5.4, 6.1, 6.2, 6.3, 6.4, 6.5, 7.1, 7.2, 7.3, 7.5, 8.2, 8.7, 9.8, 13.4, 13.5, 13.6, 14.5, 15.1, 15.3, 16.2, 16.5, 16.8, 17.1, 17.2, 17.3, 17.4, 17.6, 19.1, 21.3, 22.2, 22.3, 22.4, 23.2, 23.5, 23.6, 23.7, 23.8, 24.1, 24.4, 24.6, 24.7, 25.2, 25.3, 25.5, 25.6, 26.5, 26.10, 26.11, 26.12, 26.13, 27.1, 27.3, 27.6, 27.10, 28.2, 29.1, 29.2, 29.12, 30.2, 30.3, 30.4, TN.1

1.1: Carolee Winstein
1.2: Melinda Sue Gordon. © Warner Bros.
1.3: Tyler Ott
1.4: Rosie Draper
2.1: NASA, N. Benitez (JHU), T. Broadhurst (Racah Institute of Physics/The Hebrew University), H. Ford (JHU), M. Clampin (STScI), G. Hartig (STScI), G. Illingworth (UCO/Lick Observatory), the ACS Science Team, and ESA
2.2: Adam Evans, www.sky-candy.ca
2.3: Courtesy of NASA/SDO and the AIA, EVE, and HMI science teams
2.6: Property of the estate of Matthew H. Zimet. Courtesy Eva Zimet
2.7: © Best View Stock/Alamy
2.8: Image of Earth: NASA
2.9: © Picture Press/Alamy
2.10: © Russell Kightley/Science Source
2.11: Image of Earth: NASA
3.1: Waldseemuller map: map image courtesy of the Norman B. Leventhal Map Center at the Boston Public Library/Sidney R. Knafel Collection at Phillips Academy, Andover, MA. Ortelius map: from Library of Congress Geography and Map Division Washington, D.C. Bowen map: Geographicus Rare Antique Maps
3.3: Double Negative Visual Effects: Eugénie von Tunzelmann and Oliver James
3.6: Kip Thorne. © Warner Bros.
4.2: United States Government, adapted by Kip Thorne
4.5: © Lia Halloran, www.liahalloran.com
4.6: © Lia Halloran and Kip Thorne
4.7: © Lia Halloran and Kip Thorne
5.5: Courtesy NASA/JPL-Caltech
5.6: Double Negative Visual Effects: Eugénie von Tunzelmann and Oliver James
5.7: Karl Schwarzschild: photograph by Robert Bein, courtesy AIP Emilio Segrè Visual Archives. Roy Kerr: Roy P. Kerr. Stephen Hawking: © Richard M. Diaz. Robert Oppenheimer: en.wikipedia.org—J._Robert_Oppenheimer. Andrea Ghez: Mary Watkins (UCLA)

5.8: Keck/UCLA Galactic Center Group; Andrea Ghez
5.9: Akira Fujii/ESA/Hubble
6.5: © Kip Thorne, patterned after illustrations by Edward Teo (2003)
7.4: Left: © Robert Gendler (robgendlerastropics.com). Right: © Kip Thorne
7.6: Steve Drasco, assistant professor of physics, California Polytechnic State University, San Luis Obispo
7.7: Courtesy NASA/JPL-Caltech
8.1: Double Negative Visual Effects: Eugénie von Tunzelmann and Oliver James
8.3: Star field image: Alain Riazuelo, IAP/UPMC/CNRS; drawing: Kip Thorne. Movie camera icon courtesy basarugur
8.4: Star field image: Alain Riazuelo, IAP/UPMC/CNRS; drawing: Kip Thorne
8.5: Image: Double Negative Visual Effects: Eugénie von Tunzelmann and Oliver James; drawing: Kip Thorne. © Warner Bros. and Kip Thorne
8.6: Image: Double Negative Visual Effects: Eugénie von Tunzelmann and Oliver James; drawing: Kip Thorne. © Warner Bros. and Kip Thorne. Movie camera icon courtesy basarugur
9.1: Photo: NASA/STScI. Spectrum: Maarten Schmidt
9.2: Property of the estate of Matthew H. Zimet. Courtesy Eva Zimet
9.3: Property of the estate of Matthew H. Zimet. Courtesy Eva Zimet
9.4: Property of the estate of Matthew H. Zimet. Courtesy Eva Zimet
9.5: James Guillochon
9.6: James Guillochon
9.7: Double Negative Visual Effects: Eugénie von Tunzelmann and Oliver James
9.8: Movie camera icon courtesy basarugur. © Kip Thorne
9.9: Double Negative Visual Effects: Eugénie von Tunzelmann and Oliver James
10.1: Steve Drasco, assistant professor of physics, California Polytechnic State University, San Luis Obispo
11.2: NASA Goddard Earth Sciences Data and Information Services Center/Giovanni. Special thanks to James G. Acker.
13.1: © Kip Thorne and Richard Powell, www.atlasoftheuniverse.com
13.2: Freeman Dyson
13.3: Copyright © American Institute of Aeronautics and Astronomics, Inc. 1983. All rights reserved.
14.1: Apple © Preto Perola/Shutterstock.com. Ant: © Katarzyna Cielecka/Fotalia.com
14.2: Wormhole drawing on left previously published in Misner, Thorne, and Wheeler (1973); remainder of figure: Kip Thorne
14.3: Property of the estate of Matthew H. Zimet. Courtesy Eva Zimet
14.4: Property of the estate of Matthew H. Zimet. Courtesy Eva Zimet
14.6: Left: © Catherine MacBride. Right: © Mark Interrante
14.7: Property of the estate of Matthew H. Zimet. Courtesy Eva Zimet
14.8: Property of the estate of Matthew H. Zimet. Courtesy Eva Zimet
15.2: Left: Kip Thorne. Right: Double Negative Visual Effects: Eugénie von Tunzelmann and Oliver James. © Warner Bros. and Kip Thorne
15.4: Left: Kip Thorne. Right: Double Negative Visual Effects: Eugénie von Tunzelmann and Oliver James. © Warner Bros. and Kip Thorne
16.1: Photos courtesy of the LIGO Laboratory
16.3: Courtesy of the LIGO Laboratory
16.4: Images produced by Robert McGehee, Ian MacCormack, Amin Nikbin, and Keara Soloway from a simulation by F. Foucart et al. (2011)
16.6: © Lia Halloran, www.liahalloran.com
16.7: Image courtesy Robert Owen, Department of Physics and Astronomy, Oberlin College
16.9: Figure courtesy SXS collaboration, www.black-holes.org
16.10: Steffen Richter
17.7: Top: © Xinhua/Xinhua Press/Corbis. Bottom: © AFLO/Nippon News/Corbis

17.8: NASA/JPL/University of Arizona
18.1: Data: Huan Yang, Aaron Zimmerman; image: Oliver James and Eugénie von Tunzelmann. © Warner Bros.
21.1: Courtesy NASA/JPL-Caltech
21.2: Left: courtesy John Schwarz. Right: Steve Jennings/Getty Images
22.1: History of Science Collection, John Hay Library, Brown University
23.1: Image of Sun: Courtesy of NASA/SDO and the AIA, EVE, and HMI science teams.
23.4: Randall photo: © Tsar Fedorsky, 2014. Sundrum photo: Raman Sundrum, professor of theoretical physics, University of Maryland, College Park
24.2: NASA/JPL-Caltech/GSFC/SDSS
24.3: Andrew Fruchter (STScI) et al., WFPC2, HST, NASA
24.8: NASA/JPL
25.1: Melinda Sue Gordon. © Warner Bros.
25.4: ESA–GOCE High Level Processing Facility
25.7: Kip Thorne. © Warner Bros.
25.8: Melinda Sue Gordon. © Warner Bros.
26.1: Released into the public domain by PoorLeno
26.2: Photo courtesy of the LIGO Laboratory
26.3: Property of the estate of Matthew H. Zimet. Courtesy Eva Zimet
26.4: ChrisVanLennepPhoto/Shutterstock.com
26.6: © Stephen Hawking, John Preskill, Kip Thorne
26.7: Left: © Matthew W. Choptuik 2000. Middle and right: © Kip Thorne
26.8: Photo by Irene Fertik 1997
26.9: © Lia Halloran, www.liahalloran.com
27.2: Kip Thorne for drawing; Double Negative for image of *Endurance*. © Kip Thorne and Warner Bros.
27.4: Jeff Darling, www.diseno-art.com
27.5: Kip Thorne for drawing; Double Negative for image of *Endurance*. © Kip Thorne and Warner Bros.
27.7: Kip Thorne for drawing; Double Negative for image of *Endurance*. © Kip Thorne and Warner Bros.
27.9: Kip Thorne for drawing; Double Negative for image of *Endurance*. © Kip Thorne and Warner Bros.
28.1: Kip Thorne for drawing; Double Negative for image of Endurance. © Kip Thorne and Warner Bros.
28.4: Drawing © Kip Thorne. Floating man adapted from illustration by Cameron D. Bennett
29.3: Tesseract: "Hypercube." Licensed under Creative Commons Attribution-Share Alike 3.0 via Wikimedia Commons—http://commons.wikimedia.org/wiki/File:Hypercube.svg#mediaviewer/File:Hypercube.svg, adapted by Kip Thorne. Floating man: illustration by Cameron D. Bennett
29.4: Image of stars and galaxies: my distortion of Figure 2.1, which is courtesy NASA, N. Benitez (JHU), T. Broadhurst (Racah Institute of Physics/The Hebrew University), H. Ford (JHU), M. Clampin (STScI), G. Hartig (STScI), G. Illingworth (UCO/Lick Observatory), the ACS Science Team, and ESA. Shadow of floating man is adapted from illustration by Cameron D. Bennett
29.5: Drawing © Kip Thorne. Floating man adapted from illustration by Cameron D. Bennett. Girl adapted from illustration by Kamenetskiy Konstantin/Shutterstock.com
29.6: Tesseract: "Hypercube." Licensed under Creative Commons Attribution-Share Alike 3.0 via Wikimedia Commons—http://commons.wikimedia.org/wiki/File:Hypercube.svg#mediaviewer/File:Hypercube.svg, adapted by Kip Thorne. Floating man: illustration by Cameron D. Bennett. Standing girl: Kamenetskiy Konstantin/Shutterstock.com. Lines, arrows, and text: Kip Thorne
29.7: Drawing © Kip Thorne. Floating man: Cameron D. Bennett. Standing girl: Kamenetskiy Konstantin/Shutterstock.com
29.9: Drawing © Kip Thorne. Floating man: Cameron D. Bennett. Standing girl: Kamenetskiy Konstantin/Shutterstock.com

29.10: Drawing © Kip Thorne. Floating man: Cameron D. Bennett. Standing girl: Kamenetskiy Konstantin/Shutterstock.com

29.11: © Christopher Nolan

29.13: Drawing © Kip Thorne. Floating man: Cameron D. Bennett

30.5: Drawing © Kip Thorne. Shadow of floating man is adapted from an illustration by Cameron D. Bennett

30.6: M. C. Escher's *Waterfall* © 2014 The M. C. Escher Company–The Netherlands. All rights reserved

BIBLIOGRAPHY

Abbott, E. A. (1884). *Flatland* (Dover Thrift Edition 1992, New York); widely available on the web, for example at http://en.wikisource.org/wiki/Flatland_(second_edition).

Adkins, J. F., Ingersoll, A. P., and Pasquero, C. (2005). "Rapid Climate Change and Conditional Instability of the Glacial Deep Ocean from the Thermobaric Effect and Geothermal Heating," *Quaternary Science Reviews*, **24**, 581–594.

Al-Khalili, J. (2012). *Black Holes, Wormholes, and Time Machines*, 2nd edition (CRC Press, Boca Raton, Florida).

Barrow, J. D. (2011). *The Book of Universes: Exploring the Limits of the Cosmos* (W. W. Norton, New York).

Bartusiak, M. (2000). *Einstein's Unfinished Symphony: Listening to the Sounds of Space-Time* (The Berkeley Publishing Group, New York).

Baum, R., and Sheehan, W. (1997). *In Search of the Planet Vulcan: The Ghost in Newton's Clockwork Universe* (Plenum Trade, New York).

Baumgart, A., Jennerjahn, T., Mohtadi, M., and Hebbeln, D. (2010). "Distribution and Burial of Organic Carbon in Sediments from the Indian Ocean Upwelling Region Off Java and Sumatra, Indonesia," *Deep-Sea Research I*, **57**, 458–467.

Begelman, M., and Rees, M. (2009). *Gravity's Fatal Attraction: Black Holes in the Universe*, 2nd edition (Cambridge University Press, Cambridge, England).

Billings, L. (2013). *Five Billion Years of Solitude: The Search for Life Among the Stars* (Penguin Group, New York).

Brans, C. (2010). "Varying Newton's Constant: A Personal History of Scalar-Tensor Theories," *Einstein Online*, 1002; available at http://www.einstein-online.info/spotlights/scalar-tensor.

Carroll, S. (2011). *From Eternity to Here: The Quest for the Ultimate Theory of Time* (Oneworld Publications, Oxford, England).

Carroll, S. (2014). *Preposterous Universe*, http://www.preposterousuniverse.com/blog/.

Choptuik, M. W. (1993). "Universality and Scaling in Gravitational Collapse of a Massless Scalar Field," *Physical Review Letters*, **70**, 9.

Davies, P.C.W., and Brown, J. R. (1986). *The Ghost in the Atom: A Discussion of the Mysteries of Quantum Physics* (Cambridge University Press, Cambridge, England).

Dyson, F. J. (1963). "Gravitational Machines," in *Interstellar Communication*, edited by A.G.W. Cameron (W. A. Benjamin, New York), pp. 115–120.

Dyson, F. J. (1968). "Interstellar Transport," *Physics Today*, October 1968, pp. 41–45.

Ehlinger, L. (2007). *Flatland: The Film*, currently available on YouTube at http://www.youtube.com/watch?v=eyuNrm4VK2w; see also http://www.flatlandthefilm.com.

Emerson, S., and Hedges, J. I. (1988). "Processes Controlling the Organic Carbon Content of Open Ocean Sediments," *Paleoceanography*, **3**, 621–634.

Everett, A., and Roman, T. (2012). *Time Travel and Warp Drives* (University of Chicago Press, Chicago).

Feynman, R. (1965). *The Character of Physical Law* (British Broadcasting System, London); paperback edition (MIT Press, Cambridge, Massachusetts).

Forward, R. (1962). "Pluto—the Gateway to the Stars," *Missiles and Rockets*, **10**, 26–28.

Forward, R. (1984). "Roundtrip Interstellar Travel Using Laser-Pushed Lightsales," *Journal of Spacecraft and Rockets*, **21**, 187–195.

Foucart, F., Duez, M. D., Kidder, L. E., and Teukolsky, S. A., "Black Hole–Neutron Star Mergers: Effects of the Orientation of the Black Hole Spin," *Physical Review* D **83**, 024005 (2011); also available at http:arXiv:1007.4203.

Freeze, K. (2014). *The Cosmic Cocktail: Three Parts Dark Matter* (Princeton University Press, Princeton, New Jersey).

Gamow, G. (1947). *One, Two, Three . . . Infinity: Facts and Speculations of Science* (Viking Press, New York; now available from Dover Publications, Mineola, New York).

Gregory, R., Rubakov, V. A., and Sibiryakov, S. M. (2000). "Opening Up Extra Dimensions at Ultra-Large Scales," *Physical Review Letters*, **84**, 5928–5931; available at http://lanl.arxiv.org/abs/hep-th/0002072v2.

Greene, B. (2003). *The Elegant Universe: Superstrings, Hidden Dimensions, and the Quest for the Ultimate Theory*, 2nd edition (W. W. Norton, New York).

Greene, B. (2004). *The Fabric of the Cosmos: Space, Time, and the Texture of Reality* (Alfred A. Knopf, New York).

Guillochon, J., Ramirez-Ruiz, E., Rosswog, S., and Kasen, D. (2009). "Three-Dimensional Simulations of Tidally Disrupted Solar-Type Stars and the Observational Signatures of Shock Breakout," *Astrophysical Journal*, **705**, 844–853.

Guth, A. (1997). *The Inflationary Universe* (Perseus, New York).

Hartle, J. (2003): *Gravity: An Introduction to Einstein's General Relativity* (Pearson, Upper Saddle River, New Jersey).

Hawking, S. (1988). *A Brief History of Time: From the Big Bang to Black Holes* (Bantam Books, New York).

Hawking, S. (2001). *The Universe in a Nutshell* (Bantam Books, New York).

Hawking, S., and Mlodinow, L. (2010). *The Grand Design* (Bantam Books, New York).

Hawking, S., Novikov, I., Thorne, K. S., Ferris, T., Lightman, A., and Price, R. (2002). *The Future of Spacetime* (W. W. Norton, New York).

Hawking, S., and Penrose, R. (1996). *The Nature of Space and Time* (Princeton University Press, Princeton, New Jersey).

Hedges, J. I., and Keil, R. G. (1995). "Sedimentary Organic Matter Preservation: An Assessment and Speculative Synthesis," *Marine Chemistry,* **49**, 81–115.

Isaacson, W. (2007). *Einstein: His Life and Universe* (Simon & Schuster, New York).

Kennefick, D. (2007). *Traveling at the Speed of Thought: Einstein and the Quest for Gravitational Waves* (Princeton University Press, Princeton, New Jersey).

Lemonick, M. (2012). *Mirror Earth: The Search for Our Planet's Twin* (Walker, New York).

L'Engle, M. (1962). *A Wrinkle in Time* (Farrar, Strauss and Giroux, New York).

Levin, J., and Perez-Giz, G. (2008). "A Periodic Table for Black Hole Orbits," *Physical Review D,* **77**, 103005.

Lynden-Bell, D. (1969). "Galactic Nuclei as Collapsed Old Quasars," *Nature,* **223**, 690–694.

Maartens, R., and Koyama K. (2010). "Brane-World Gravity," *Living Reviews in Relativity* **13**, 5; available at http://relativity.livingreviews.org/Articles/lrr-2010-5/.

Marolf, D., and Ori, A. (2013). "Outgoing Gravitational Shock-Wave at the Inner Horizon: The Late-Time Limit of Black Hole Interiors," *Physical Review D,* **86**, 124026.

McKinney, J. C., Tchekhovskoy, A., and Blandford, R. D. (2012). "Alignment of Magnetized Accretion Disks and Relativistic Jets with Spinning Black Holes," *Science,* **339**, 49–52; also available at http://arxiv.org/pdf/1211.3651v1.pdf.

McMullen, C. (2008). *The Visual Guide to Extra Dimensions.* Volume 1: *Visualizing the Fourth Dimension, Higher-Dimensional Polytopes, and Curved Hypersurfaces* (Custom Books).

Meier, D. L. (2012). *Black Hole Astrophysics: The Engine Paradigm* (Springer Verlag, Berlin).

Merritt D. (2013). *Dynamics and Evolution of Galactic Nuclei* (Princeton University Press, Princeton, New Jersey).

Misner, C. W., Thorne, K. S., and Wheeler, J. A. (1973). *Gravitation* (W. H. Freeman, San Francisco).

Nahin, P. J. (1999). *Time Machines: Time Travel in Physics, Metaphysics and Science Fiction,* 2nd edition (Springer Verlag, New York).

Obst, L. (1996). *Hello, He Lied: & Other Truths from the Hollywood Trenches* (Little, Brown, Boston).

Obst, L. (2013). *Sleepless in Hollywood: Tales from the New Abnormal in the Movie Business* (Simon & Schuster, New York).

O'Neill, G. K. (1978). *The High Frontier: Human Colonies in Space* (William Morrow, New York; 3rd edition published by Apogee Books, 2000).

Paczynski, B., and Wiita, P. J. (1980). "Thick Accretion Disks and Supercritical Luminosities," *Astronomy and Astrophysics,* **88**, 23–31.

Pais, A. (1982). *"Subtle Is the Lord . . .": The Science and the Life of Albert Einstein* (Oxford University Press, Oxford, England).

Panek, R. (2011). *The 4% Universe: Dark Matter, Dark Energy, and the Race to Discover the Rest of Reality* (Houghton Mifflin Harcourt, New York).

Penrose, R. (2004). *The Road to Reality: A Complete Guide to the Laws of the Universe* (Alfred A. Knopf, New York).

Perryman, M. (2011). *The Exoplanet Handbook* (Cambridge University Press, Cambridge, England).

Pineault, S., and Roeder, R. C. (1977a). "Applications of Geometrical Optics to the Kerr Metric. I. Analytical Results," *Astrophysical Journal*, **212**, 541–549.

Pineault, S., and Roeder, R. C. (1977b). "Applications of Geometrical Optics to the Kerr Metric. II. Numerical Results," *Astrophysical Journal*, **213**, 548–557.

Poisson, E., and Israel, W. (1990). "Internal Structure of Black Holes," *Physical Review D*, **41**, 1796–1809.

Randall, L. (2006). *Warped Passages: Unraveling the Mysteries of the Universe's Hidden Dimensions* (HarperCollins, New York).

Rees, M., ed. (2005). *Universe: The Definitive Visual Guide* (Dorling Kindersley, New York).

Roseveare, N. T. (1982). *Mercury's Perihelion from Le Verriere to Einstein* (Oxford University Press, Oxford, England).

Schutz, B. (2003). *Gravity from the Ground Up: An Introductory Guide to Gravity and General Relativity* (Cambridge University Press, Cambridge, England)

Schutz, B. (2009). *A First Course in General Relativity*, 2nd edition (Cambridge University Press, Cambridge, England).

Singh, P. S. (2004). *Big Bang: The Origin of the Universe* (HarperCollins, New York).

Shostak, S. (2009). *Confessions of an Alien Hunter: A Scientist's Search for Extraterrestrial Intelligence* (National Geographic, Washington, DC).

Stewart, I. (2002). *The Annotated Flatland: A Romance of Many Dimensions* (Basic Books, New York).

Teo, E. (2003). "Spherical Photon Orbits Around a Kerr Black Hole," *General Relativity and Gravitation*, **35**, 1909–1926; available at http://www.physics.nus.edu.sg/~phyteoe/kerr/paper.pdf.

Thorne, K. S. (1994). *Black Holes & Time Warps: Einstein's Outrageous Legacy* (W. W. Norton, New York).

Thorne, K. S. (2002). "Spacetime Warps and the Quantum World: Speculations About the Future," in Hawking et. al. (2002).

Thorne, K. S. (2003). "Warping Spacetime," in *The Future of Theoretical Physics and Cosmology: Celebrating Stephen Hawking's 60th Birthday*, edited by G. W. Gibbons, S. J. Rankin, and E.P.S. Shellard (Cambridge University Press, Cambridge, England), Chapter 5, pp. 74–104.

Toomey, D. (2007). *The New Time Travelers: A Journey to the Frontiers of Physics* (W. W. Norton, New York).

Visser, M. (1995). *Lorentzian Wormholes: From Einstein to Hawking* (American Institute of Physics, Woodbury, New York).

Vilenkin, A. (2006). *Many Worlds in One: The Search for Other Universes* (Hill and Wang, New York).

Wheeler, J. A., and Ford, K.(1998). *Geons, Black Holes and Quantum Foam: A Life in Physics* (W. W. Norton, New York).

Will, C. M. (1993). *Was Einstein Right? Putting General Relativity to the Test* (Basic Books, New York).

Witten, E. (2000). "The Cosmological Constant from the Viewpoint of String Theory," available at http://arxiv.org/abs/hep-ph/0002297.

Yang, H., Zimmerman, A., Zenginoglu, A., Zhang, F., Berti, E., and Chen, Y. (2013). "Quasinormal Modes of Nearly Extremal Kerr Spacetimes: Spectrum Bifurcation and Power-Law Ringdown," *Physical Review D*, **88**, 044047.

INDEX OF PEOPLE

MOVIE CHARACTERS

REAL PEOPLE

INDEX OF SUBJECTS

Page numbers in *italics* refer to illustrations.